Airborne Doppler Radar: Applications, Theory, and Philosophy

Airborne Doppler Radar: Applications, Theory, and Philosophy

by
Martin Schetzen
Northeastern University
Boston, Massachusetts

Volume 215
PROGRESS IN
ASTRONAUTICS AND AERONAUTICS

Frank K. Lu, Editor-in-Chief
University of Texas at Arlington
Arlington, Texas

Published by the
American Institute of Aeronautics and Astronautics, Inc.
1801 Alexander Bell Drive, Reston, Virginia 20191-4344

American Institute of Aeronautics and Astronautics, Inc., Reston, Virginia

1 2 3 4 5

ISBN 1-56347-828-5

Table of Contents

Preface

THIS TEXT is based, in part, on my work as a consultant for the Apollo 11 project, the project to land and return the first human being on the moon. An onboard computer was used to land on the moon. The external data required by the computer were the vehicle velocity and altitude above the lunar surface. These data were obtained from an onboard Doppler radar. However, there was a concern that the radar performance over the moon might differ from that over the Earth. To allay this concern, I was requested to review all radar theory to determine if such differences could exist. My conclusion was that the radar should perform over the moon exactly as over the Earth. However, possible differences between the lunar surface and the Earth's surface concerned me. The vehicle velocity relative to the surface was to be obtained from the Doppler radar spectrum. In my review of radar theory, I found that no exact theory for the Doppler spectrum had been developed. Instead, an approximation in which a Doppler frequency is assigned to each point on the surface was used. There was a paradox called the "smooth earth paradox" associated with this approximation. A smooth earth is one in which the surface is smooth like a mirror. There should be no change in the radar echo as the radar travels across such a surface, so in contradistinction to the approximation, the velocity should not be determinable. This paradox alone indicated to me that a proper Doppler theory must include the reflecting surface as a parameter that affects the Doppler spectrum. Because it appeared to me that the reflecting surface should affect the spectrum, I was concerned that differences between the lunar surface and the Earth's surface could result in a difference in the Doppler radar spectrum. Some lunar surface data had been obtained by telescopic observation and from the data telemetered back to earth from the Pioneer series of rockets, but the data were not sufficient to allay my concerns. I thus set about working on the development of an exact theory for the Doppler spectrum.

The theory I developed resolved the smooth earth paradox and allayed my concerns. In guidance and control applications, the actual shape of the Doppler spectrum is not really necessary; only its center frequency and bandwidth are required. Unfortunately my theory required the determination of a graph of the actual Doppler spectrum from which these spectrum parameters could be determined. Each graphical point required the evaluation of a few integrals by a computer. This limited the desirableness of my theory even though it was experimentally shown to be correct and used for the Lunar Lander (the Eagle). For this, I received the Apollo Achievement Award and the Apollo Certificate of Achievement. After the success of Apollo 11, the Apollo research groups that had developed it were disbanded. Even so, I continued to extend my theory in my spare time in order to make it more useful. Equations from which the center frequency and bandwidth can be obtained without first determining the actual Doppler spectrum were developed and an important result obtained from these equations is what I call the *Airborne Doppler uncertainty principle*.

This principle states that in airborne Doppler radar there is a lower limit of the product of position uncertainty and velocity uncertainty. This extension is included in this text.

In writing this text, I took the opportunity to learn, for the first time, something about Doppler's life and works. I was struck by how modern was his scientific approach. I discuss this in the context of my personal scientific philosophy in Chapter 1. The Doppler effect is of great scientific importance because of its applicability to many scientific areas. I discuss some of these applications and their importance in Chapter 2. Our main concern is the use of the Doppler effect in aircraft applications. I thus briefly discuss how the Doppler effect can be used for aircraft flight control in Chapter 3. The accuracy of the control is determined by the Doppler spectrum bandwidth. This motivates a discussion in Chapter 4 of my quasi-static approximation of the Doppler spectrum, which is the classic approximation in which I have included the effect of backscattering from the surface. The development of my exact theory of the airborne Doppler spectrum utilizes concepts from generalized harmonic analysis. To establish common notation and viewpoint, a précis of the required elements of harmonic and system analysis is included in Chapter 5. This is followed by Chapters 6 and 7 in which my Doppler theory and *Airborne Doppler uncertainty principle* are developed and discussed. Most of the material in these chapters is contained in an Apollo report.[1] My extension of this theory is presented in Chapter 8. In it, specific equations for just the Doppler frequency and the Doppler spectrum bandwidth are developed and expressed in coordinates relative to the antenna. Chapter 9 is a numerical study of these results and a comparison with the results obtained from the quasi-static approximation. The design of a Doppler laser radar is discussed in section 9.4, where the quasi-static approximation is shown to result in significant approximation errors. This example is an illustration of the *Doppler uncertainty principle*.

A computer program was written for the numerical studies of the derived equations in Chapter 9. For this, a new two-dimensional integration algorithm was developed and is described in Chapter 10. This chapter also includes a description of the computer program and its use contained on the included disk. The antenna pattern and the backscattering function used by program on the disk are the ones described in the text. However, other antenna patterns and backscattering functions can be used by modifying the program as described in section 10.3. Thus, without recourse to the theoretical development in the text, the program disk can be used by radar designers and experimentalists to determine the effect of the various antenna and backscatter parameters on the center frequency and bandwidth of the Doppler spectrum.

I am grateful to Paul Zarchan, editor of the AIAA Press, for inviting me to write this text that gave me the motivation to do the numerical studies included in Chapter 9. These studies could not have been made without Dr. William Jennings, who wrote the computer program for the numerical studies. During

---

[1] Schetzen, M., *The Power Density Spectrum of the Echo from an Airborne Doppler Radar*, Apollo Guidance, Navigation, and Control Report R-541, Massachusetts Instrumentation Laboratory, March, 1966.

the writing of this text, I became professor emeritus at Northeastern University. I am grateful to Vice Provost Pat Meserve, who located an office for me in which I could complete this text. Finally, I thank my wife, Jeannine, without whom this text probably would not have been completed. She made important textual suggestions and continued to encourage me to persevere in spite of many obstacles.

Martin Schetzen
Brookline, Massachusetts
August 1, 2004

Chapter 1

Scientific View

I. Scientific Reality

IN WRITING this text of my work on the echo spectrum of an airborne Doppler radar, I took the opportunity to read, for the first time, Doppler's original paper in which he develops and discusses his theory of this apparent frequency shift. I was struck by how modern was his scientific approach. Scientific explanations at the time were objectively based. In essence, Doppler's scientific approach was subjectively based in that he considered physical reality as only what is experimentally observed.

Classically, a basis for the acceptance of a scientific theory was a satisfying logical argument; observation and measurement were not considered of primary importance. For example, Plato (427?–347 B.C.E.) has Socrates (469?–394 B.C.E.) state in his discussion with Simmias in the *Phaedo*,[1]

> "Well then, what about the actual getting of wisdom? Is the body in the way or not, if a man takes it with him as companion in the search? I mean, for example, is there any truth for men in their sight or hearing? Or so poets are forever dinning in our ears, do we hear nothing and see nothing exactly? Yet, if these of our bodily senses are not exact and clear, the others will hardly be, for all inferior to these, don't you think so?
>
> "Certainly," Simmias said.
>
> "Then," said Socrates, "when does the soul get hold of the truth? For whenever the soul tries to examine anything in company with the body, it is plain that it is deceived by it."
>
> "Yes."
>
> "And I suppose it reasons best when none of the senses disturbs it, hearing and sight, or pain, or pleasure indeed, but when it is completely by itself and says good-bye to the body, and so far as possible has no dealings with it, when it reaches out and grasps that which really is."
>
> "That is true."

---

[1] Rouse, W.H.D., *Great Dialogues of Plato*, The New American Library Mentor Book, 1956, pp. 468–469.

"And is it not then that the philosopher's soul chiefly holds the body cheap and escapes from it, while it seeks to be by itself?"

"So it seems."

......

"And would he do that most purely who should approach each with his intelligence alone, not adding sight to intelligence, or dragging in any other sense along with reasoning, but using the intelligence uncontaminated alone by itself, while he tries to hunt out each essence uncontaminated, keeping clear of eyes and ears and, one might say, of the whole body, because he thinks the body disturbs him and hinders the soul from getting possession of truth and wisdom when body and soul are companions—is this not the man, Simmias, if anyone, who will hit reality?"

"Nothing could be more true, Socrates," said Simmias.

It was not respectable for a member of the Greek aristocratic class to do manual labor and so such experimental work was, at the time, left to the artisan class. Because there was little concern with doing experiments to validate a theory, Greek science at the time was mostly speculative. One consequence was that science did not become a real force in Greek society and it became part of the liberal studies for the autocratic class. Science thus became a subject of contemplation and ceased to be a means of transforming the conditions of life. Even such established professions as the architect and the physician were only quasi-respectable. The experts involved in such professions engaged in it only to the extent to which they could be regarded as the possessors of purely theoretical knowledge that they used to direct the labor of others.

It was Copernicus (1473–1543) who was one of the first champions of a movement to use experimental observations as the basis for validating scientific theories. To Copernicus, observation was a view of objective reality and a scientific theory was a description of that reality. As time progressed, the scientific concept of reality changed. The best brief statement of the modern view of reality was by Max Planck (1858–1947), who stated in 1932 that there are two fundamental principles of scientific philosophy:[2]

1) There is a real outer world which exists independent of our attempt to know it.
2) The real outer world is not directly knowable.

That is, scientific philosophy today is that a scientific theory is only a subjective model of experimental results. I thus was struck by the following statement in Doppler's paper that was first published in 1842:[3]

... it seems remarkable to me that in science of both light and sound, as well as in the general theory of waves, one has—at least as far as I know—failed to take into consideration one commonly occurring circumstance! It seems that the fact has been ignored that in speaking of light and sound waves as

---

[2] Planck, M. *Where is Science Going?* W.W. Norton & Co., 1932, p. 82.

[3] Doppler, C. A., *Über das farbige Licht der Doppelsterne und einiger anderer Gestirne des Himmels* [On the Colored Light of Double Stars and Certain Other Stars of the Heavens], Reprinted in Eden, A., *The Search for Christian Doppler*, Springer-Verlag, Wien, New York, 1992.

being the causes of light and sound perception and not merely as objective processes, one should not so much ask in what space of time and with what level of intensity the creation of waves actually takes place but rather at which time intervals and with what strength these ether waves are picked up and perceived by the eye or ear of the observer. It is from these purely subjective conditions, and not from the objective facts that the color and intensity of light-perception or the pitch and volume of any sound are dependent. If by any chance a numerical difference should occur between the objective processes and the subjective results, then one must without any doubt adhere to the subjective evaluation. At first glance it may well appear as though what has been said above is of merely academic distinction rather than observations that are accompanied by practical consequences. On this point the honored reader, as soon as he has given the following lines his worthy consideration, shall decide for himself. So long as it is assumed that the observer as well as the source of the waves remain in their original position without moving, there can be no doubt that the subjective evaluation will numerically coincide completely with the objective. What, however, if either the observer or the source—or even both together—change their position, move away from or nearer to each other, and this with a speed comparable to that with which the waves are propagated? Could also in this case such a conformity be reckoned with? I hardly believe that the reader could feel inclined to give an affirmative answer to this question without previous investigation! In fact, nothing seems easier to comprehend than that the distance and time interval between two successive waves must become shorter for an observer who is hurrying towards the oncoming waves and longer if he is moving away, and similarly, in the first case the intensity of the wave is stronger and in the second it must necessarily decrease. With a movement of the source of the waves itself, a similar change naturally takes place in the same sense. We know from general experience that a ship with a moderately deep draught which is steering towards the oncoming waves has to receive, in the same amount of time, more waves with a greater impact compared with a ship that is not moving or is even traveling along on the direction of the waves. If this should be valid for waves of water, then should it not also be applied with necessary modifications to air and ether waves? It hardly seems that anything of consequence could be raised against this! Under the circumstances it might seem expedient to set out the very simple formulae that are necessary for this and, by experimentally applying them to sound waves, we think that we could at the same time be of some small service to acoustics.

His approach is that the only meaningful characteristics of a wave are those subjectively observed and not some objective description. That is, Doppler's argument effectively is that physical reality is only what is experimentally observed. This also is the underlying basis of my model of Doppler radar return discussed in this text.

The concepts of reality in science, philosophy, and theology are not entirely the same. The misunderstanding of the difference has often led to public disputes on topics that are, in large part, based on different definitions of reality. To explain the difference, I'll begin with a brief discussion of the modern scientific philosophy of observation and of the meaning of the resulting scientific models that are formed. Scientific reality together with the scientific models that we create and the influence these models have on our thinking will be discussed. As we shall see, models lie at the base of all our scientific work, whether theoretical or experimental.

II. Scientific Models

The fundamental activity of modern science is the development and analysis of models. In theoretical work, one attempts to analyze a model of some phenomenon. In experimental work, one tries to verify an aspect of a model. In engineering work, one tries to exploit some aspect of a model for practical applications. All technical work is based on some scientific model. Even though every thought we have, whether scientific or not, is based on some model, their interaction with and their influence on our thinking are not often discussed.

To begin, a scientific model in modern scientific thought is just a representation of a phenomenon. A model is desired because it can be used to predict the outcome of experiments involving the phenomenon and we often can obtain new insights concerning the phenomenon from the model. It is important to note that we are only interested in the prediction of experimental results obtained with the model. If the predicted results obtained from the model do not correspond with the actual experimental results, we modify the model to make them correspond. For example, the model of the electron has been modified many times. At first, it was a negative charge in orbit about the nucleus of an atom. However, to explain the emission spectra observed, the orbits were quantized; to explain other experiments, the electron was given spin and the spin was quantized. The electron was then given a wave motion with a wavelength as a result of deBroglie's theory in 1925, which was experimentally verified by Davisson and Germer in 1927. The model of the electron is being refined continually.

Does the electron really exist? Such a question is not within the domain of science. Science is not at all concerned whether the model actually represents objective reality because it is the model that constitutes scientific reality. The common concepts of objective reality fall within the domain of the philosopher or the theologian, not the scientist. The only concern of science is whether the predictions obtained from the model correspond with actual experimental results. We thus can say that a scientific model mimics reality. All we then can say of the electron model is that experiments proceed *as if* the electron exists. In this sense, the model is scientific reality.

As another example, consider the model of light. At first, the model was that of a wave as suggested by Huygens (1629–1695). However, Newton (1642–1727) believed that the phenomenon of double refraction could be explained better with a model of light as a stream of corpuscles, but he never developed his model. Later, Fresnel (1788–1827) was able to explain double refraction with the wave model and so the theory of light in the nineteenth century was constructed using Huygen's wave model rather than Newton's corpuscular model. However, the photoelectric effect could not be explained by the wave model. Therefore Einstein (1879–1955) resurrected the corpuscular model in which the corpuscles were called photons. Thus there were two models for light, one a wave model, and the other a photon model. The wave model could be used for the analysis of such phenomena as diffraction but not for other phenomena such as the photoelectric effect. On the other hand, the photon model could be used to explain such phenomena as the photoelectric effect but not diffraction. This duality of models was resolved by the wave-particle model of deBroglie (1892–1987) in which the model of light is that of a stream of photons that travel as a wave which, according to the complementarity principle, only one—the particle or the wave manifestation—can be observed in a given experiment. deBroglie essentially combined the wave model and the photon model to form his wave-particle model. We do not say that this is the true structure of light; we only say that experiments with light proceed in accordance with this model.

A model then only has scientific meaning in terms of experimental outcomes. Thus, I may ask if the desk in front of me is really there. As a scientist, all I can do is perform experiments. I may shine a light on it. I then would say that my visual sensation is in accordance with the model of the desk being there. Similarly, I may put my hand on it. I then say that my tactile sensation is in accordance with the model of the desk being there. Is it really there? As a scientist, I can only say that experimental results correspond the model of the desk being there.

Correspondingly, the question of existence in which there are no experimental outcomes is a meaningless scientific question. For example, let's say that a question arose as to whether there is an angel sitting on my desk. I would begin by saying that I cannot see the angel. Of course, you would respond, angels are invisible. I then would put out my hand and say that I cannot feel it. Your response would be that you can not feel the angel because angels are not material. I then would argue that there is no experimental method by which I can detect the presence of the angel. You would agree. Thus, I would conclude that the question of whether there is an angel sitting on my desk is scientifically meaningless.

The question of the presence of an angel may have meaning in certain modes of philosophy or theology, but not scientifically. For example, consider the story in the bible of the meeting of Abraham with the three angels on the plains of Mamre or the story of the combat of Jacob with the angel at Peniel on the banks of the Jabbok. The actual existence of the angels in the stories is secondary to the teaching of moral behavior and the human condition. To convey these teachings, the bible often speaks in images such as "the still, small voice" that the prophet Elijah hears on Mount Horeb or, after Cain murders his brother Abel, God telling him that he hears the "voice of your brother's blood crying from the earth."

In telling of an occurrence, the actual events involved were often modified somewhat and images were inserted in order to teach a lesson or enhance the image of an individual, group, or country. Thucydides ($\sim$460–$\sim$400 B.C.E.) was one of the first to write a history in which accuracy of relating events was the main objective. In his work *The History of the Peloponnesian War*, which was fought from 431 to 404 B.C.E., his main concerns were not philosophical or theological but rather historical accuracy, and it is often considered to be one of the first of such works.

There are many philosophic approaches used for the development of a philosophy of existence. In contrast to the scientific approach that uses sense data as

I have discussed, one philosophic approach uses the philosophic principle of Descartes (1596–1650), *Cogito ergo sum*, I think therefore I exist. Note that this approach allows the existence of an isolated individual to be defined much like the creature in Pointland, the Abyss of No dimensions, described in *Flatland* by E.A. Abbott. The creature, which is a point living in the land of zero dimensions, is heard to say "What It (meaning itself) thinks, that It utters; and what It utters, that It hears; and It itself is Thinker, Utterer, Hearer, Thought, Word, Audition; it is the One, and yet the All in All. Ah, the happiness ah, the happiness of Being!" This is solipsism. The religious approach, however, defines existence in terms of connections. That is, the existence of an individual in religious philosophy gains definition only in terms of the individual's connections to others. These connections are used in religion as the basis for its development of ethics and morality.

Please note that questions concerning reality as discussed in these examples are meaningless in science. Such questions are in the province of the philosopher and/or the theologian, not the scientist.

III. Model Determination

Now, a problem with a scientific model of a phenomenon is that it is obtained via measurements, but the measurements do not necessarily uniquely describe the phenomenon. Thus different models may predict the same experimental results. How then do we choose between such models? First, consider the model for planetary motion proposed in the year 140 by Ptolemy (127–151) versus the model proposed in 1543 by Copernicus (1473–1543). I'm certain you are familiar with the great controversy that raged when, in 1632, Galileo (1564–1642) published the results of his observations that, he argued, confirmed the Copernican model of planetary motion and disclaimed the Ptolemy model. This controversy ended by Galileo being required to publicly renounce his conclusions even though he privately still believed them to be correct. Even so, was Ptolemy really wrong and Copernicus correct? The essential difference between the two proposed models of planetary motion is that Ptolemy used the earth as the origin of his coordinate system while Copernicus used the sun as the origin of his coordinate system. Therefore, the two proposed models of planetary motion differed only by a translation of the coordinates and so were equivalent! Why then was the Copernican model considered to be the correct one? The important difference is that the paths of the planets are epicycloids in the Ptolemy geocentric model while they are circles in the Copernican heliocentric model.

There is an old principle used in science called Occam's razor. The principle is named after a British scholastic philosopher, William of Occam (1285–1347). He insisted that an explanation should be as simple as possible. His actual statement, called Occam's razor, is "Entities [that is, concepts, assumptions, etc.] are not to be multiplied beyond necessity." He used this principle of economy of thought in all his philosophical works and so popularized it that his name became attached to it. Relative to our discussion, Occam's razor is that among competing models, the most simple model is the one to choose.

Thus the Copernican model of planetary motion was chosen over the Ptolemy model beacuse the circle of the Copernican model is a simpler curve than the epicycloid of the Ptolemy model. This choice did not, at the time, invalidate the Ptolemy model. It was just that the Copernican model was a simpler model of the planetary orbits.

There is a fundamental reason for choosing the simplest model. A simpler model is so desirable because the simpler the model, the more easily our limited minds can grasp the essence of the phenomenon being modeled. This grasp often results in a new view of the phenomenon from which we conceive of new concepts basic to the phenomenon. For example, Johannes Kepler (1571–1630) was motivated to study planetary motion using the Copernican model. For this study, he joined Tycho Brahe's astrological group in order to eventually appropriate the precise measurements his group had made. With these data in hand, Kepler was able to deduce his three empirical laws. We use the diagram below depicting the elliptical orbit of a planet moving about the sun to show the three laws.

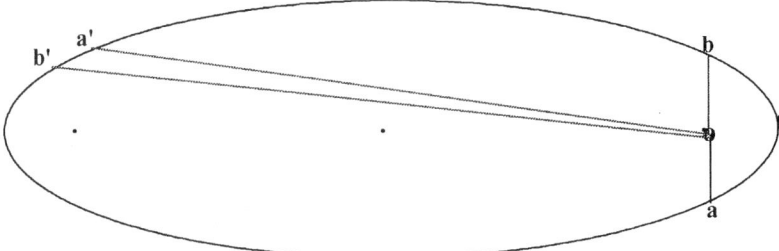

1) The planets move in elliptical orbits with the sun at one focus. This radically modified the Copernican model of circular orbits.

2) The line joining the planet with the sun sweeps out equal areas in equal times. Thus the time it takes the planet to travel from a to b is the same as the time it takes it to travel from a' to b'. This is the reason a comet travels so much faster when it is near our sun and takes so long to make one orbit.

3) The third was published in 1619, which was 10 years after he published the first two. The square of the period of a planet about the sun is proportional to the cube of its distance from the sun.

In fact, it was the third empirical law of Kepler that Newton used to deduce the inverse-square law of gravitation that he published in his *Principia* in 1687. Thus, using Tycho Brahe's data, the simplicity of the Copernican model enabled Kepler to deduce his laws of planetary motion. Newton then used Kepler's third law to deduce his model of gravitation.

Interestingly, the simplicity and success of the Copernican model has so captivated our thought process that, when considering the solar system, we view the sun as the center of the coordinate system but when we shift our thoughts to, for example, a satellite of the planet Saturn, we immediately change our frame of reference and view Saturn as the center of the coordinate system. We thus observe in this case how our thought process has been so prejudiced by the Copernican model that we even have difficulty in viewing phenomena in terms of any other model.

Another interesting illustration is the model of light. In proposing his wave model of light, Huygens wrote in his 1690 publication, *Traité de la Lumiére*, "[Light] spreads as sound does, by spherical surfaces and waves: for I call them waves from their resemblance to those which are seen to be formed in water when a stone is thrown into it." Because water waves have water as a medium through which to travel and light travels through a vacuum, Huygens decided that the propagation of light was caused by "particles of the aether . . . a substance . . . approaching to perfect hardness and possessing a springiness . . . in which the movement of light is successively communicated."

In fact, as Doppler states in my quote from his paper, this is the model of light waves he used. This model so took hold that there was a significant effort to experimentally demonstrate the existence of Huygens's "aether." The famous Michelson–Morley experiment in 1887 was one such experiment in which it was shown, as is not the case with water waves, that the velocity of light is a constant irrespective of the velocity of the observer. That is, if you are moving while measuring the velocity of a water wave, then the velocity you would measure is the sum of the water wave velocity and your velocity. With light, however, your measurement would result in the same velocity irrespective of your velocity. The confirmation of this experimental result puzzled scientists and a number of models were proposed. The favored model was the one proposed by Fitzgerald (1851–1901) and by Lorentz (1853–1928). In this model, by some unknown mechanism, motion through the ether results in distances being shortened in the direction of motion. The shortening is by an amount that results in the measured velocity of light being exactly the same irrespective of the velocity of the observer. This shortening was called the Lorentz–Fitzgerald contraction. Fitzgerald believed that there was a conspiracy in nature to prevent us from detecting motion through the ether. Einstein was rather bothered by this conspiracy concept.

IV.   Observability and Existence

I stated earlier that the question of existence in which there are no experimental outcomes is a meaningless question in science. Note that experimental observations can only reveal the relation between entities. The Michelson–Morley experiment demonstrated that the ether, like my example of the angel, is not observable. Thus, the result that the ether cannot be observed renders it a meaningless scientific entity and so it should be removed from scientific models.

A major source of revolutions in intellectual thought is those who break loose from the prejudices of accepted models and are able to form a new model along a new paradigm that is not only consistent with previous results but also with the new observed results. In business, this is often called "thinking outside the box." An excellent example is the revolutionary model Einstein proposed to include the result of the Michelson–Morley experiment. This model, which is his special theory of relativity, was obtained by going back to basic principles and asking the very basic question, "What do we mean by simultaneity?" That is, if the velocity of light is a constant irrespective of the observer's velocity, what do we mean when we say that two events occur simultaneously? This is the fundamental question Einstein posed for himself as the basis for the development of his model of relativity. One constraint that he imposed was that his new model include the Newton laws as a special case. In this manner, his new model was consistent with previous experimental results as well as the new experimental results obtained by Michelson and Morley. As we all know, one prediction of this model is the famous $E = mc^2$ formula that is the basis of atomic energy. Einstein further developed this model by asking simple basic questions concerning acceleration. This is his general theory of relativity in which a new model of the basic structure of space is advanced. This new model resulted in a fundamental revision of the Newtonian model of gravitation and gave explanation to many puzzling aspects of Newton's model. For example, one puzzling aspect was why the gravitational mass of a body is exactly equal to its inertial mass. The gravitational mass is the value of mass one uses when saying that the attractive force between two bodies is proportional to the product of their masses. The inertial mass is the value of mass one uses when saying that force equals mass times acceleration. In the Newtonian model, there was no reason why the gravitational and inertial masses should be equal. However, their equality is a logical consequence of the Einstein model of gravitation. New models such as the Copernican model of planetary motion and the Einstein model of gravitation often make certain observations appear logical that, in the old model, seemed to be ad hoc. In some cases, the attempt to "explain" some observation that is ad hoc in the accepted model results in a radically new model in which the observation appears to be logical. Such a new radical model is a restructuring of scientific reality with an attendant major scientific advance. Einstein's model of gravity, his general theory of relativity, is one such example.

Such revolutionary thinking is difficult to support financially because it often requires a great deal of time and effort without producing tangible results. In consequence, innovative intellectual pursuits in the past have been accommodated mainly in the universities. Although universities declare their commitment to the encouragement and support of scholarship, even its minimal assistance in many universities today is often subordinate to short-term business concerns and organization politics. Consequently, fundamental scholarship often is done by highly motivated individuals at significant personal sacrifice and so many innovative ideas lay fallow these days. If science is to progress, methods for the proper ascertainment of promising scholarships and giving them an opportunity to develop are basic problems with which we must come to grips. Otherwise, like the poor farmer, we are eating our seed.

The development of a fundamentally new model sometimes also results in major scientific breakthroughs. Thus Einstein's new model of space and time caused a revolution in cosmology. Edwin Hubble (1889–1953) was an astronomer who did "battleship" astronomy. I call his experimental observational studies battleship astronomy because the telescope projecting from the spherical dome of the observatory is, to me, reminiscent of a battleship. In 1929, Hubble made a series of careful measurements at the Mount Wilson Observatory. From these measurements, he concluded that the stars are receding from earth at a velocity that is proportional to their distance. This observation is at the basis of the big-bang theory because this is what one would observe if there were an initial "big bang." At any instant after an explosion, a particle observed twice as far from the explosion as another must have been traveling twice as fast as the other particle. Thus, the velocity of a particle is proportional to its distance from the center of the explosion. However, because our solar system is not at the center of the universe, we then should also observe some stars approaching us, yet no such stars have been observed. This is a problem with the Hubble model. In accordance with Einstein's new model of space and time, we now reinterpret Hubble's observed result by saying that the stars only appear to be traveling away from us because space itself is expanding. A way to visualize this is to consider a metal sheet with many white dots painted on its surface and one red dot in their midst. Now uniformly heat the metal sheet. As it expands, no matter where a white dot is relative to the red dot, the distance between it and the red dot will increase at a velocity proportional to the distance separating them. This is an example of space expanding. This new model is not only consistent with Hubble's observations but also explains why no stars appear to be approaching us.

I began by stating that a scientific model is just a representation of a phenomenon that is used to predict the outcome of experiments and from which we often can obtain new insights. For this, we saw that Occam's razor was invoked in order that the simplest model is the one chosen among competing models. This has the advantage of enabling our limited human minds to grasp the essence of the phenomenon being modeled. Even so, the simplest model often is still too complicated for us and so we need to further simplify it. There are a number of ways this is done and each way has its consequences. One technique is to change the questions we ask of the phenomenon.

V. Stochastic Models

Instead of asking for specific values, the model can be simplified by only asking for certain average values. This is accomplished by creating a statistical model. It is important to note that the statistical model so formed does not necessarily imply that the phenomenon is not deterministic. The phenomenon may well be deterministic but, for convenience in our theoretical considerations, we choose to create a statistical model and so model the phenomenon as being random.

For example, when a coin is tossed, we say that the probability of the coin landing with the head side up is one-half. However, the experiment is really deterministic because whether the coin will land heads up can be determined if the initial conditions were known. You may have practiced tossing a coin so that it would almost always land heads up. Of course, you learned by practice to toss the coin so that each toss had essentially the same initial conditions. However, it is difficult if not impossible to know the exact initial conditions for each toss. Thus, we often model the coin toss as random by assuming that the initial conditions will vary from toss to toss. Also, for simplicity, we usually assume that the initial conditions vary from toss to toss in such a manner that a head or a tail on any toss is equally likely to occur and is independent of whether it was head or tail on any other toss. This model is called an ideal coin or an honest coin. Clearly, on the average, 50 percent of the tosses of an ideal coin will land heads up. *Randomness in this example, we note, is in the eye of the beholder.* The classical definition of the probability of an event is that it is equal to the ratio of the number of equally likely ways an event can occur in an experiment divided by the number of equally likely experimental outcomes. The probability of obtaining a head on a toss of an ideal coin thus is $1/2$. Interestingly, Feller states, "It has also been claimed that the probabilities $1/2$ are due to experience. As a matter of fact, whenever refined statistical methods have been used to check on actual coin tossing, the result has been invariably that head and tail are not equally likely."[4] Even so, we stay with our model of an ideal coin even though no ideal coins have been observed. We preserve this model of an ideal coin not only for its simplicity, but also for its usefulness. If you have had a basic course on probability, just consider how many situations were modeled in terms of the tossing of an ideal coin.

A stochastic model is obtained by attributing a randomness to certain parameters of the phenomenon. The resulting stochastic model is then amenable to analysis from which we often can obtain new insights concerning the phenomenon. As a simple illustration of this, consider how we model a gas-filled balloon. The gas consists of many gas molecules bumping into each other as they move about. The balloon pressure is due to the molecules bouncing off the balloon surface. One approach to determine the pressure is to write the differential equation for the motion of each gas molecule within the balloon and try to solve the resulting set of many millions of simultaneous differential equations. The solution requires knowledge of the initial conditions, which are the position and velocity of every gas molecule at one instant of time. The knowledge cannot be obtained experimentally. The reason is not because of the difficulty of performing the experiment but rather because this knowledge is theoretically impossible to obtain because, in accordance with the Heisenberg uncertainty principle, the more accurate the position of a particle is known, the less accurate can the particle velocity be known at that instant. The uncertainty in knowing the initial conditions results in uncertainty of the molecular positions and velocities that grows with time. However, even without the theoretical problems involved with this approach, we can easily see that this approach is not practical because of the enormous number of equations that would have to be solved simultaneously. Thus, rather than ask for the pressure as a function of time and position on the balloon surface, we only ask for certain average values of the pressure. To determine the averages, we model the gas as a very large collection of gas molecules moving randomly within the balloon. With this model we only need ask for certain statistical properties of the gas molecules. For example, the classical result that the average balloon volume times the average pressure on the balloon surface is proportional to the absolute temperature of the gas within the balloon can be derived in this manner. The specific motion of each molecule then is no longer of concern. In fact, it is outside the scope of the stochastic model. The development of this approach is one of the great contributions of Willard Gibbs (1839–1903).

Another example of how a stochastic model is formed is the Doppler radar theory I developed for the Apollo 11 lunar lander radar. A computer was used

---

[4] W. Feller, *An Introduction to Probability Theory and its Applications*, J. Wiley & Sons, Second Edition, 1957, p. 19.

to guide and land the lunar lander on the moon. For the landing, the height of the lunar lander above the moon surface was determined by a radar altimeter that transmits a pulse to the surface and measures the time for the reflected pulse to be received. The velocity was determined by measuring the Doppler shift. The Doppler shift is proportional to the relative velocity between the lunar lander and the moon surface. A problem is that the relative velocity varies from point to point on the reflecting surface. Thus if the radar transmitted just a sinusoid with a given frequency, the echo would contain not just one frequency but a spectrum of frequencies about the transmitted frequency. Classically, the spectrum was approximated by assigning a Doppler frequency to each point on the surface and calculating a spectrum by weighting the frequencies with the antenna pattern. Although this simple model results in a reasonable approximation of the spectrum under certain conditions, we shall see later in this text that it does not in others. Because the computer was to use this data to land the vehicle, a more reliably accurate model was desired. Additionally, detailed knowledge of the moon surface was not available. A view of the general structure of the surface in some local areas was obtained in 1964 from photographs taken of the moon surface with the Ranger spacecrafts as they crashed into the moon surface. Because detailed knowledge of the moon surface was not available, it was not certain that the Doppler spectrum obtained experimentally on earth would be the same as that one would obtain on the moon. An exact theory would include the effect of the surface characteristics on the spectrum. With an exact theory, the surface characteristics that affect the spectrum and their effect on the spectrum would be known. Experiments then could be performed to determine those surface characteristics.

Because the spectrum is, as we shall see, obtained from a certain average of the received waveform, I formed a stochastic model of the reflected wave from the terrain. For this I used a subterfuge, which was a *gedanken experiment*—a thought experiment. As I'll discuss later in this text, I assumed that the experiment of obtaining the reflected wave is repeated many times; in fact, I imagined the experiment is repeated an infinite number of times. Each waveform obtained is then considered to be a member of an ensemble of such waveforms. Such an ensemble of waveforms is called a stochastic process. The Doppler spectrum then can be calculated from certain statistical properties of the ensemble. These statistical properties are affected by certain statistical properties of the terrain. I thus formed a statistical model of the terrain that only requires knowledge of certain average properties of the terrain. By working out the theory in this manner, I was able to obtain explicit expressions for the Doppler spectrum in terms of the antenna pattern and certain statistical properties of the terrain. This model then showed which properties of the terrain affect the spectrum and the manner in which it is affected.

When a phenomenon is too complicated because it contains too many elements for reasonable analysis or the exact parameters are not known, we often simplify the model, as in the examples just mentioned, by creating a stochastic model and asking of it only certain averages. It is important to note that the stochastic model so formed does not necessarily imply that the phenomenon is not deterministic. The phenomenon may well be deterministic. However, for convenience in our theoretical considerations, we chose to create a stochastic model.

As we saw, the stochastic model is obtained by attributing a randomness to certain parameters of the phenomenon. A basic scientific concern is whether there is a basic uncertainty in nature. That is, does every stochastic model result from incomplete knowledge of the physical phenomenon, or is there a fundamental vagueness in natural processes? Max Born (1882–1920) and Werner Heisenberg (1901–1976) thought there is. On the other hand, Albert Einstein (1879–1955) and Max Planck (1858–1947) expressed the opinion that, as illustrated in my examples, our knowledge of physical phenomena is incomplete and the introduction of new concepts may result in deterministic models.

For example, the accepted model of radioactive decay is a stochastic model in accordance with the Born–Heisenberg view. However, is radioactive decay truly stochastic, or is it stochastic because we do not understand the mechanism of radioactive decay well enough? It appears to me that it may be possible to construct a deterministic model in accordance with the Einstein–Planck view by using bifurcation theory. This a mathematical theory of equations whose solution bifurcates at a critical point so that two or more solutions are possible near a given set of conditions. A simple example that illustrates bifurcation is a canoe. We've all been told never to stand up in a canoe because standing up raises the center of gravity and will destabilize the canoe. Why? Well, when we sit in the canoe, we can rock the canoe by a large angle without it rolling over in the water. As the center of gravity is raised, this maximum angle decreases until, at a certain height, the maximum angle is zero. At that point, the slightest motion will cause the canoe to roll over. This is the bifurcation point at which there are two solutions, one being the canoe rolling over to the right and the other being the canoe rolling over to the left. Thus a deterministic model for radioactive decay could be constructed if one can show that there is a bifurcation point about the conditions at which radioactive disintegration takes place. Slight variations within the atom thus can result in its radioactive disintegration. For such a model, it would not be possible to predict when a particular atom will disintegrate for reasons similar to my coin example in which it is not possible to predict whether the toss of a coin will result in a head. Also, as with the tossing of a coin, the radioactive emission would appear to be random, with the number of atoms disintegrating being proportional to the number of atoms that have not yet decayed. This is what is observed experimentally. I propose this model only to show that it may be possible to construct a deterministic model of radioactive decay instead of the presently accepted stochastic model.

VI. Meaningless Questions

Often, a model is complicated because the elements of which it is constructed are themselves rather complicated. In electric circuits, for example, a simple wire resistor is not just an element that we often model as an ideal resistor. In reality, the time-varying current flowing in the wire is accompanied by a time-varying magnetic field about the wire that results in a back emf. One way of accounting for this is to include an ideal inductance, called lead inductance, in series with the ideal resistor. In addition, there is a time-varying electric field across the resistor that results in a displacement current. This is modeled by including an ideal capacitance, called stray capacitance, across the ideal resistor. Already we can see that the model of the simple wire resistor is becoming rather complicated, and these are just first-order corrections. At low frequencies, the simple ideal resistor results in a sufficiently accurate model. At higher frequencies, the lead inductance and the stray capacitance must be included in the model. At higher frequencies yet, one must use a more complicated model of the wire resistor that accounts for time-delay and other effects. The simplest model consistent with analysis objectives is used. The model, such as that of a circuit, is thus composed of ideal elements such as ideal resistors, inductors, capacitors, voltage sources, etc. The construction of a model using ideal elements often means that there are certain questions I cannot ask of the model because they are meaningless in the context of a model composed of ideal elements. The attempt to answer such meaningless questions often ends with a seeming paradox. The literature is replete with such paradoxes. I'll illustrate a few of the more common paradoxes of this sort.

One of the simplest to discuss is "What happens if an irresistible force meets an immovable object?" This question defines two ideal worlds: a world in which there is an irresistible force and an ideal world in which there is an immovable object. By definition, an immovable object cannot exist in the world that contains the irresistible force. Similarly, an irresistible force cannot exist in the world that contains an immovable object. Consequently, the two worlds must be disjoint and so cannot overlap. We thus cannot ask what happens if the irresistible force meets the immovable object because that implies that the two worlds overlap. The question is not one for which there is no answer. Rather, the question itself is meaningless!

A similar situation is illustrated by the following simple circuit consisting of two ideal batteries.

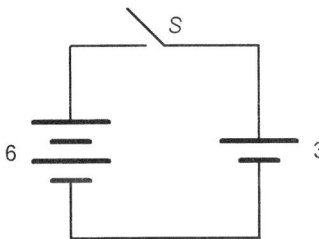

What happens when the switch, S, is closed? This simple example lies at the basis of some common circuit "paradoxes." Many of my students say that infinite current flows if the switch is closed. However, by definition, the voltage across an ideal battery is the same irrespective of what is connected across it. Thus the voltage across the 3- and 6-volt ideal batteries must stay the same if the switch is closed. To say that the switch is closed consequently states that $3 = 6$. Thus it is not what happens if the switch is closed but rather the switch cannot be closed! Closing the switch is inconsistent with the model. The question of what happens when the switch is closed thus is meaningless. Note the similarity of this example with the previous one. Of course, you clearly could physically connect two real batteries as in the figure and close the switch. The difference then is that you used real batteries and not ideal ones. Real batteries have some internal resistance. A good model of a real battery is an ideal one in series with an ideal resistor whose value is equal to the internal resistance of the real battery. The switch now can be closed because the voltage difference across the switch is made zero by the voltage drop across these resistors.

The model of a feedback system offers an example of another type of paradox that can arise if a physical system is not properly modeled. For this example, consider the model of a feedback system as shown below.

The equations of this model are

$$y(t) = x(t) + z(t)$$

and

$$z(t) = Ky(t)$$

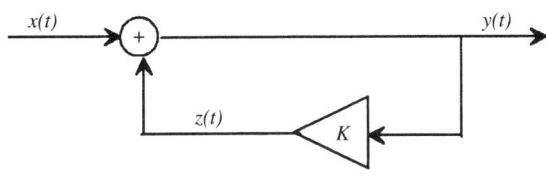

The solution of these two equations for the output $y(t)$ is

$$y(t) = \frac{1}{1-K} x(t)$$

For the special case in which $K = 2$, we have from this result that for a unit step input $x(t) = u(t)$, the response is $y(t) = -u(t)$. Although this indeed is a solution of the system equations of the model, it is one that the physical system never exhibits! The reason is that there always is some delay in any physical system due to the fact that the size of any component of the physical system d is greater than zero. Because no wave can travel faster than the speed of light c, there must be a delay of at least d/c seconds between the input and the output of any physical component. This delay could be exceedingly small but not zero. For example, if the component size is about 3 cm, there must be a delay of about 10^{-10} seconds $= 0.1$ nanoseconds. We thus must be careful when modeling a component with zero delay because the solution of the system model then may not be consistent with the solution of the physical one. The solution of a system model with zero delay must always be obtained as the limit as the component delays go to zero or, equivalently, with the component delays being infinitesimally small. A proper model of any physical system must thus include some delay.

To understand the effect of a small delay, let us examine the model in which a small delay t_0 is inserted in the feedback path as shown below.

The response $y(t)$ of this system for $K = 2$ and the input $x(t) = u(t)$ is a staircase function with the value $2^n - 1$ in the time interval $(n - 1)\, t_0 < t < nt_0$. With this result, we can examine the solution as the delay time t_0 tends to zero. The following table is a list of values of the system output at $t = 1/3$ microsecond for various values of t_0.

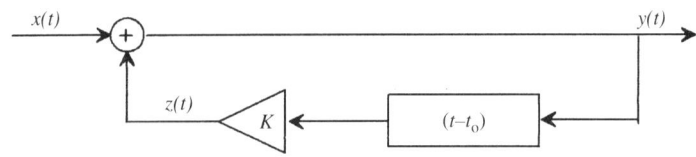

t_0 (microseconds)	t/t_0	n	$y\left(\dfrac{10^{-6}}{3}\right) = 2^n - 1$
10^{-6}	$\dfrac{1}{3}$	1	1
10^{-7}	$\dfrac{10}{3}$	4	15
10^{-8}	$\dfrac{10^2}{3}$	34	$1.7 \cdot 10^{10}$
10^{-9}	$\dfrac{10^3}{3}$	334	$3.5 \cdot 10^{100}$
10^{-10}	$\dfrac{10^4}{3}$	3,334	$4.3 \cdot 10^{1,003}$

The output amplitude at $t = 1/3$ microsecond is seen to become arbitrarily large as the delay time t_0 becomes arbitrarily small. In fact, you should note that this is true for any given value of t because for any given value of t, n increases as t_0 decreases. We thus observe that, for our system, the proper solution for zero delay is that $y(t)$ is infinite. Generally, the solution of any system model with zero delay must always be obtained as the limit as the component delays go to zero or, equivalently, with the component delays being infinitesimally small.

These examples illustrate the care that must be taken in modeling a physical system when certain component values are modeled with zero value. We always must interpret zero as the limit of the solution as the component value tends to zero or, equivalently, the solution obtained by using an infinitesimal value for the component value. Similarly, problems can arise by considering an impulse as having zero width. To avoid problems with an impulse, it should always be considered to have an infinitesimal width.

Whenever one attempts to simplify analysis by replacing model components with ideal elements, the questions one asks of the simplified model should be examined to make certain they are meaningful. As an example, for analytic simplicity in analyzing communication systems, a signal is often modeled as being bandlimited. This is a reasonable model for many communication problems. However, for the determination of an optimum prediction of a signal, it may not be a good approximation. The reason is that a bandlimited signal is a member of a class of waveforms that are quasi-analytic. This means that the future values of the waveform can be determined by a causal system with arbitrarily small error. For example, if your speech waveform were bandlimited, the past of your speech waveform can be used to predict with arbitrarily small error what you will say at any time in the future. That would imply that you do not have the free will to say what you desire. The whole philosophy of ethics requires the existence of free will because, without it, no one is responsible for their actions. So, if you believe in free will, you must conclude that your speech waveform is not bandlimited. That means that the power density spectrum of your speech waveform is not zero even in the megahertz and gigahertz frequency range. The spectrum is rather small up there, but it is just that small amount that results in the minimum theoretical error in predicting what you will say to grow with increasing time in the future for which the prediction is made. From this, you can note that a waveform that contains information cannot be bandlimited because if it were bandlimited, its future values can be predicted and so there is no need to receive it. An information-bearing waveform is a waveform for which new information is inserted at various times. Thus it must be a waveform whose future cannot be predicted and so it cannot be bandlimited. The bandlimited approximation often used is thus a good approximation only if the analysis does not make essential use of the fact that the waveform spectrum is zero beyond a certain frequency. That is, as with my previous examples, we must make certain that the questions we ask of a model are consistent with the idealizations we made in constructing the model.

With the ever increasing speed, memory, and parallelism of computers, they can be used to simulate ever increasing complex models. Already computers are used to simulate models of the atmosphere for weather prediction and to simulate certain models of biologic processes. An accurate analytical analysis of these models has not been possible. With the computer, new techniques are being developed to obtain an acceptably accurate analysis. This has resulted in more accurate weather prediction. Also we are obtaining a better understanding of certain biologic processes that may result in the opening of new medical techniques never previously imagined. A good example is the genome project in which the genetic structure is being determined.

With the computer, we now can analyze models for which there are no reasonable analytic methods available for their analysis and so we can ask questions of the model that we could not ask previously. For example, the computer can be used to simulate and so study models of chaotic systems, neural networks, and self-organizing systems. The computer can be used not just to plot curves but also to study certain dynamic characteristics of the system. A good example is in the study of self-organizing systems. In all these studies, however, is critically important that the model be carefully studied to make certain that it is not an improper model in the sense I've discussed.

VII. Scientific Mysticism

Without the careful development of scientific models, there is a regression into various forms of mysticism that call themselves scientific be it in astrology, evolution, medicine, or cosmology. The case of Immanuel Velikovsky (1895–1979) offers an interesting example in cosmology. He published a book in 1950 titled *Worlds in Collision*. He argued in his book that many miraculous events described in the bible could be shown to be scientifically valid by modeling them as catastrophic events in the solar system. This book created quite a controversy in the public arena as well as in the scientific community. Eventually, however, his model was rejected as a scientific model. The reason for the rejection is that a scientific model must contain the elements I've discussed and Velikovsky's model did not contain those elements.

As I mentioned earlier, the model proposed by Fitzgerald to explain the negative result of the Michelson–Morley experiment was mystically based because it implied that there is a conspiracy in nature to prevent us from detecting motion through the ether. It thus was not acceptable as a scientific model. A proper scientific model cannot be based on an act of God such as a miracle or a catastrophic event that appears preposterous within the larger fabric of science. Einstein's relativity model was accepted because it contains no mysticism and is consistent with all previous experimental results. Velikovsky's model was rejected by the scientific community as a scientific model because it required catastrophic events that appear preposterous within the larger fabric of science.

VIII. Scientific Education

The development and proper analysis of scientific models also require individuals like Doppler who think in original and innovative ways. For this, it is critical that students be intellectually stimulated and broadly educated with an emphasis on fundamental concepts. Certainly it is important that technical courses present practical applications to illustrate a concept and to provide a more insightful understanding and appreciation of possible uses of the fundamental concepts discussed. However, it is not necessary to teach the ever changing latest technology; that will be learned quickly on the job if the student is involved with the technology after graduation. Thus, technical education should be broad with an emphasis on fundamental concepts to stimulate original and innovative thinking that is so necessary for our further scientific and technological development. What are required for technical and scientific advancement are individuals who have an appreciation of the large range of scientific and technical possibilities. For this, the presentation of many of the concrete concepts in technical education should include highlights illuminating their infinite varieties of interpretation.

Additionally, progress today has been intimately mated with technology. The world also is faced with an evolving technology that it cannot stop. The broad social consequences of many technical developments require the technically educated individual to consider these consequences simultaneous with the development of the new technology. Thus a technical education also requires a basic component of liberal education. The liberal education component will assist in the aesthetic growth of the student with a comprehension and appreciation of the variety of values. This requires the simultaneous consideration of many aspects of life. This differs from the abstract thought developed in the scientific courses in which one often subdivides considerations of various consequences. The liberal education component thus should be designed to help educate the student for the increasingly important simultaneous consideration of the technology and its social consequences.

A curriculum such as the one I propose would provide the student with the basic elements of thought necessary for the consideration of problems in proper context and scientific modes of thought for problem analysis. We should be careful not to turn out students as parts of a machine because it may be for a machine that is obsolete. Curricula that produce problem solvers are producing machine parts. We require curricula that produce students who are intellectually enthusiastic about their subjects and are eclectic in their approach in order to create the new needed machines. If not, we fossilize intellectually, incompatible with society that is a dynamic organic whole.

Chapter 2

Doppler Effect

T HE DOPPLER effect is named after Christian Andreas Doppler, who was the first to analyze it. The effect is an apparent shift of the frequency of a wave due to the relative velocity between the observer and the source of the wave. Anyone who has stood by a railroad track as a speeding train passes by has experienced the Doppler effect. The Doppler effect observed is that the train whistle has a higher pitch as the train approaches and a lower pitch as the train goes away. In fact, this observation of a moving train is the first method by which Doppler's prediction of an apparent frequency shift was first experimentally verified qualitatively.

I. Christian A. Doppler

Doppler was born on 29 November 1803 in Salzburg, Austria, into a family of well-known stone masons. He had two brothers and two sisters. As was the custom, he and his brothers were expected to work in the family stone masonry business. However, Christian was sickly and so could not successfully ply the stone mason trade. It is thought that the stone dust he inhaled helped the development of consumption (tuberculosis). The family thus decided to send him to school, a fortunate decision because he became an outstanding student. His excellence eventually resulted in his being appointed in 1835 professor of elementary mathematics and commercial accounting at the State Secondary School in Prague and supplementary professor of higher mathematics and practical geometry at the Prague Technical Institute. Finally, by Imperial Decree of 6 March 1841, he was appointed full professor of elementary mathematics and practical geometry at the Polytechnic Institute in Prague. This appointment carried with it a very heavy workload, such as the reports of 800 students he had to examine and 668 written works to read and grade. (His previous position required him to lecture, examine, and grade only 400 students without assistance.) To reword the famous ditty, "The life of a professor is not an easy one." His health continued to deteriorate with this heavy burden, which was in addition to his duties as an associate member of the Royal Bohemian Society of Sciences and his theoretical research in physics. It was at a meeting of Natural Sciences Section of this society on 25 May 1842 that he read his paper entitled *On the Colored Light of Double Stars and Certain Other Stars of the Heavens* in which he first presented his ideas concerning the Doppler effect. It is interesting to note that there were only five listeners in the audience and the chairman of the session was Doppler's friend and mentor Bernard Bolzano, the well-known mathematician who was one of the first to introduce the modern concept of rigor into mathematical analysis.

The transverse wave model of light was, at the time, controversial and had not been fully accepted. Because Doppler's theory used the transverse wave model of light, he began his paper with a short review of the controversy by referencing the work of Euler and Huygens, its originators; those who finally had been convinced at the time such as Young, Fresnal, and Cauchy; those who still had their doubts such as Laplace and Poisson; and Herschel the younger who, at that time, still opposed the transverse wave model of light. Doppler then argued what has become part of the modern philosophy of science—that physical reality is only what is experimentally observed. That is, as I argue in the Chapter 1, the only true reality in science is the model of a phenomenon.

Doppler's theory was not immediately accepted. One of his main detractors was Josef Petzval, who was a professor of mathematics and mechanics at the Imperial University in Vienna. In 1850, an Institute of Physics was established at the Imperial University and Doppler was appointed its first director and professor of experimental physics in the philosophical faculty there. Petzval's attacks on Doppler appear to be more politically than scientifically motivated because, at the time of his opposition, the Doppler principle had already been confirmed through experiments. Doppler's last scientific publication, in fact, was a defense against Petzval's attacks. Doppler's health had been steadily deteriorating. He went to Vienna to convalesce and five months after his arrival, on 17 March 1853, he died. He was just 49 years old.

II.   Doppler Effect

Doppler determined the Doppler shift by using his analogy with water waves. Actually, the Doppler shift for an electromagnetic wave is not exactly the same as that for an acoustic or water wave. The reason for the difference is that acoustic and water waves require a medium through which to travel while an electromagnetic wave does not. That there is no ether through which an electromagnetic wave travels is essentially a postulate of the Einstein Theory of Relativity which, of course, was unknown in Doppler's time. However, we shall show that the Doppler shifts for acoustic and electromagnetic waves are essentially the same when the velocity of the source and the velocity of the observer is small compared with the wave velocity in the medium. In consequence, we shall see that the error in his determination of the Doppler shift is essentially zero for cases in which relativistic effects can be ignored.

To quantitatively understand the Doppler effect, consider the observation of a single frequency of the light from a star such as a spectral line of hydrogen. The single frequency waveform is the sinusoidal waveform

$$s_0(t) = A \cos[\omega_0 t + \phi] \qquad (2.1)$$

The waveform received on earth is delayed by $t_0 = r/c$ seconds in which r is the radial distance to the star and c is the velocity of light ($c \approx 186{,}300$ miles/s).

The received waveform thus is

$$s(t) = Bs_0(t - t_0) = AB\cos[\omega_0(t - t_0) + \phi]$$

$$= AB\cos\left[\omega_0\left(t - \frac{r}{c}\right) + \phi\right] \tag{2.2}$$

Now, the perceived instantaneous frequency is the rate of change of the sinusoidal phase of the received waveform, which is

$$\omega = \frac{d}{dt}\left[\omega_0\left(t - \frac{r}{c}\right) + \phi\right] = \omega_0 - \frac{\omega_0}{c}\frac{dr}{dt} \tag{2.3}$$

The perceived frequency thus differs from the emitted frequency by an amount

$$\delta = \omega - \omega_0 = -\frac{\omega_0}{c}\frac{dr}{dt} \tag{2.4}$$

The apparent frequency shift is called the Doppler shift. Note that it is proportional to the rate of change of the radial distance from the star to the earth. As shown in Fig. 2.1, d is the perpendicular distance from the earth to the line along which the star is traveling, and the star is a distance x to the left of the perpendicular line.
The radial distance then is

$$r^2 = x^2 + d^2 \tag{2.5}$$

for which

$$r\frac{dr}{dt} = x\frac{dx}{dt} \tag{2.6}$$

Now, let the star be traveling with a velocity v relative to the earth in the direction of the arrow. We then have from Eq. (2.4)

$$\frac{dr}{dt} = \frac{x\,dx}{r\,dt} = -\frac{x}{r}v = -v\sin\alpha \tag{2.7}$$

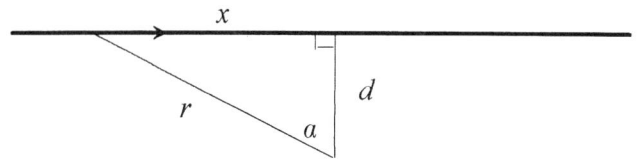

Fig. 2.1 **Geometrical representation of a moving star.**

in which α is the angle shown in Fig. 2.1. With the use of Eq. (2.4), the ratio of the Doppler shift to the emitted frequency thus is

$$\frac{\delta}{\omega_0} = -\frac{v}{c}\sin\alpha \tag{2.8}$$

Now, the approximate velocity of an electromagnetic wave in the atmosphere is $c = 3 \times 10^8$ meters/s. For a star velocity relative to the earth of 30 kilometers/s (67,108 miles/h, which is the earth's orbital velocity about the sun) we would have

$$\frac{\delta}{\omega_0} = -\frac{30 \times 10^3}{3 \times 10^8}\sin\alpha = -10^{-4}\sin\alpha \tag{2.9}$$

Figure 2.2 is a graph of the Doppler shift as a percentage of the frequency radiated from the star versus the angle α. Note that there is no Doppler shift when the star's velocity is perpendicular to the earth (for which $x = 0$) because the radial velocity of the star is zero at that point. The Doppler shift is observed to be a very small percentage of the radiated frequency. To measure this Doppler shift requires a measurement error that is only a small fraction of the Doppler shift.

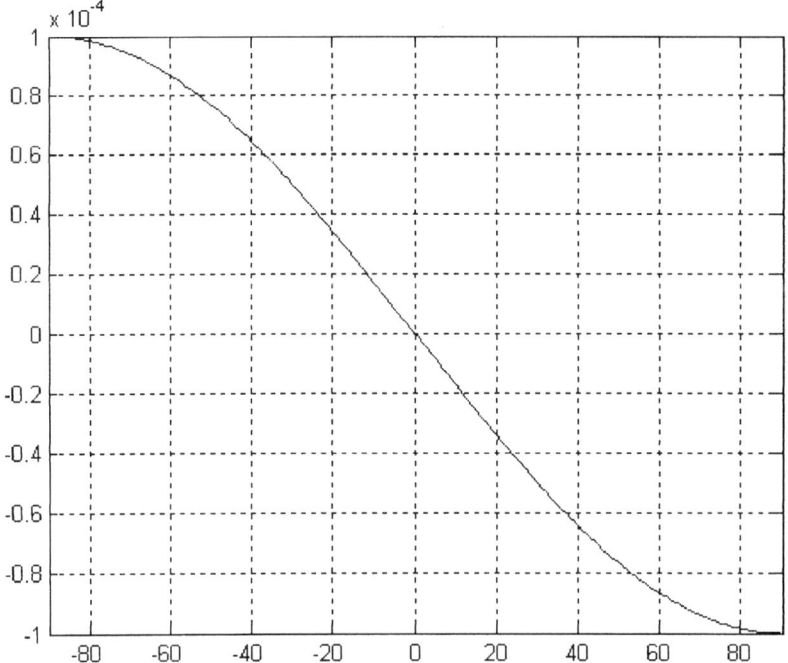

Fig. 2.2   Graph of the percentage Doppler shift vs α for the example.

Doppler had thought that the relative velocity of the stars to the earth would be so great that it would affect their apparent color and intensity. This was controversial at the time and it was later shown not to be correct. It was not such an absurd idea, though, because there was no information at the time regarding the order of magnitude of stellar velocities. Even so, the first application of Doppler's ideas was in astronomy.

A star's relative velocity was first measured by William Huggins in 1868. He measured the Doppler shift in the hydrogen F line of the light from Sirius and so calculated its relative velocity to be 47 km/s. Further measurements showed that Huggins's measurements were not very accurate. The first reasonably accurate measurement was by Hermann Vogel and Julius Scheiner who, in 1888, measured the frequency shift photographically with the use of the 11-inch Potsdam refractor. The estimated error of their measurement was 5–10 km/s. Then in 1890, James Keeler reduced the error to 1 km/s by making a visual measurement using the 36-inch refractor at the Lick Observatory. These and other measurements indicated that the spectral lines from stars in nearby galaxies were displaced downward toward the red end of the spectrum and that the more remote the galaxy, the greater is this displacement. This is the famous red shift.

The red shift was extensively examined experimentally by Edwin Hubble, an astronomer who did "battleship" astronomy. In the 1930s, Hubble made a series of careful measurements at the Mount Wilson Observatory. He announced in 1935 that his measurements of the shift of Fraunhofer calcium absorption lines of light from distant galaxies indicated that the red shift is proportional to the distance of a galaxy from us. Because the measured line displacement toward the red end of the spectrum is a negative Doppler shift, Hubble's experimental result indicated that the galaxies are moving away from us with a velocity proportional to their distance from us. This is the origin of the big-bang theory because we conclude from this that the known universe started with a big bang some 15 billion (15×10^9) years ago.[1] To understand this conclusion, consider a photograph taken of particles flying off from an explosion. The photograph is a picture of the particle positions at a given time instant after the explosion. Consequently, the distance in the photograph of a particle from the explosion center would be proportional to its velocity. Also, if this proportionality were known, the time between the explosion and the photograph could be determined.

This picture soon began to bother astronomers because it conflicted with an assumed cosmological principle that effectively requires that large-scale observations from different parts of the universe be essentially the same. This principle then implies that, no matter where in the universe the Hubble measurements were made, one should observe the same negative Doppler shift indicating that the galaxies are moving away from the observer with a velocity proportional to their distance from the observer. This would not be so with the Hubble model and is the source of the astronomers' concern. In accordance with Einstein's model of space

---

[1] Hubble's estimate was some 20 billion years ago. However, the gravitational effect of all the matter in the universe is to continually reduce the expansion rate. Taking into account the larger expansion rate in the past than that observed today results in an estimate of the big bang being about 15 billion years ago.

and time as described in his General Theory of Relativity, we now reinterpret the Hubble result by saying that the stars only appear to be traveling away from us because space itself is expanding. A way to visualize this is to consider a metal sheet with many white dots painted on its surface and one red dot in their midst. Now uniformly heat the metal sheet. As it expands, no matter where a white dot is relative to the red dot, the distance between it and the red dot will increase with a velocity proportional to the distance separating them. This is an example of space expanding. This new model is not only consistent with Hubble's observations but also with the cosmological principle. In accordance with this model of expanding space, the red shift is not produced in quite the same way as the Doppler shift. Rather, the view is that in the time light travels to us from a distant galaxy, the wavelength of the light increases due to the expansion of space. Because frequency is inversely related to wavelength, this means that the observed frequency decreases and so the observed red shift. Thus, even though Doppler's use of his theory of frequency shift to explain the differences of star color was not correct, his basic idea of frequency shift was correct and did allow the determination of star velocities that led to the big-bang theory. He is one of the fortunate ones who has been recognized for the underlying basic correctness of his theoretical idea even though his application of it was found wanting.

Cosmology, as I outlined above, was the first important application of the Doppler effect. It has, however, been used in many other important applications. The biomedical area is one of the important areas of its application. For example, in medicine it is being used in ultrasonic imaging for the determination of the flow of bodily fluids such as blood in the carotid arteries, aorta, heart, and leg veins. With Doppler sonography, one also can determine whether blood vessels sewn together during surgery are functioning properly as, for example, a bypass to an oxygen-starved area of the brain. One also can determine that the clipping off of an aneurysm in the brain has not impaired the blood flow in the vessel. In transcranial Doppler sonography, one can tell something about blood flow in the vessels without opening the skull. This is an important advance in the study of cerebrovascular disease and is important in the prediction and treatment of stroke.

However, the theory of the Doppler effect for sound is somewhat different from that for electromagnetic waves. The difference stems from the fact that the velocity of an electromagnetic wave is independent of the velocity of the observer, which is not true for sonic waves. One consequence is that there is no ether through which an electromagnetic wave travels while sonic waves require a medium through which to travel.

To illustrate the difference, consider the simple example of a source traveling at a constant velocity v_s directly toward a receiver that is stationary. If the source transmits an electromagnetic wave with a sinusoidal waveform given by Eq. (2.1), the observed received frequency will be in accordance with Eq. (2.3).

$$\omega = \omega_0 \left[1 - \frac{1}{c} \frac{dr}{dt} \right] = \omega_0 \left[1 + \frac{v_s}{c} \right] \qquad (2.10)$$

Our derivation of this result did not take into account relativistic effects because, in obtaining Eq. (2.3), we assumed $v \ll c$. If this were not so, then the relativistic contraction of time, $t = t_0\sqrt{1 - (v_s/c)^2}$, must be used so that ω_0 in Eq. (2.10) must be replaced by $\omega_0/\sqrt{1 - \beta^2}$ in which $\beta = v_s/c$. Equation (2.10) then becomes

$$\omega = \frac{\omega_0}{\sqrt{1 - \beta^2}}(1 + \beta)$$

$$= \omega_0\sqrt{\frac{1 + \beta}{1 - \beta}} \qquad (2.11)$$

If $\beta \ll 1$, we can expand the numerator and denominator of Eq. (2.11) in a power series and keep only a few terms with the result

$$\omega = \omega_0[1 + \beta]^{1/2}[1 - \beta]^{-1/2}$$

$$\approx \omega_0\left[1 + \frac{1}{2}\beta - \frac{1}{8}\beta^2\right]\left[1 + \frac{1}{2}\beta + \frac{3}{8}\beta^2\right] \qquad (2.12)$$

Keeping only terms to the second power of β, we obtain

$$\omega \approx \omega_0\left(1 + \beta + \frac{1}{2}\beta^2\right) \qquad (2.13)$$

To a first power of β, this is the nonrelativistic result given by Eq. (2.10). The nonrelativistic result is seen to be accurate within 1 percent for the determination of the Doppler shift for any velocity that is less than 14.1 percent of c ($c \approx 186,300$ miles/s). In the atmosphere, this corresponds to velocities less than 26,268 miles per second! At such a velocity, the source would travel from the east coast to the west coast of the United States in about 114 milliseconds![2]

The nonrelativistic Doppler shift for sonic waves is different than that for electromagnetic waves because sonic waves can only travel in a medium. Because sonic velocities are so much less than the velocity of light, we only need to consider the nonrelativistic theory for the Doppler shift of sonic waves. To obtain the equations for the Doppler shift, we consider the situation in which a source is traveling at a constant velocity v_s directly toward a receiver that also is traveling at a constant velocity v_r directly toward the source. The source transmits a sonic wave that, as before, is a sinusoidal wave given by Eq. (2.1). The wave that travels through the medium will contain a sequence of crests traveling through the medium at a velocity of c, which is the sonic wave velocity in the medium. For our derivation, we first consider the case in which $v_r = 0$ so that only the source is moving. Let a crest from the source be transmitted at a time $t = t_0$. It is received at the time $t_1 = t_0 + r_0/c$ in which r_0 is the distance the wave travels from the source to the receiver. Now, because the fundamental period of the sinusoid is $\tau = 1/f_0$, the next crest from the

---

[2] However, in areas such as cosmology and high energy physics, sometimes large values of β must be considered.

source will be at the time $t = t_0 + 1/f_0$. It will be received at the time $t_2 = t_0 + 1/f_0 + r/c$ in which $r = r_0 - v_s/f_0$ because the source will have traveled a distance v_s/f_0 toward the receiver. The receiver thus observes the time between crests to be

$$\Delta t = t_2 - t_1 = \frac{1}{f_0}\left[1 - \frac{v_s}{c}\right] \tag{2.14}$$

The time between crests is, of course, one period of the received sinusoid so that its frequency is

$$f_s = \frac{1}{\Delta t} = f_0\left[\frac{1}{1 - v_s/c}\right] \tag{2.15}$$

Now consider the case in which $v_s = 0$ so that only the receiver is moving. The time between crests is one period of the sinusoid, which is $\tau_0 = 1/f_0$. The distance between crests in the medium thus is $d = c\tau_0 = c/f_0$. Because the relative velocity between the wave and the receiver is $(c + v_r)$, the receiver will observe the time between crests to be

$$\Delta t = \frac{d}{c + v_r} = \frac{c}{f_0}\frac{1}{c + v_r} \tag{2.16}$$

The sinusoidal frequency observed by the receiver thus is

$$f_r = \frac{1}{\Delta t} = f_0\left[1 + \frac{v_r}{c}\right] \tag{2.17}$$

We now consider the case in which both the receiver and the source are moving. The distance between crests now is $d = c/f_s$ in which f_s is given by Eq. (2.15). Thus the sinusoidal frequency observed by the receiver is given by Eq. (2.17) in which f_0 in that equation is replaced by f_s. We then have for the case in which both the source and the receiver are traveling toward each other that the observed frequency is

$$f_r = f_0\left[\frac{1 + v_r/c}{1 - v_s/c}\right] \tag{2.18}$$

Note that the Doppler shift that results from the source velocity is different than the Doppler shift that results from the receiver velocity. If the source and receiver velocities are small compared with the wave velocity in the medium c, then by expanding the denominator of Eq. (2.18) in a power series

$$f_r \approx f_0\left[1 + \frac{v_r}{c}\right]\left[1 + \frac{v_s}{c} + \left(\frac{v_s}{c}\right)^2\right] \tag{2.19}$$

Keeping only terms to the second power we have

$$f_r \approx f_0\left[1 + \frac{v_r}{c} + \frac{v_s}{c} + \frac{v_s(v_s + v_r)}{c^2}\right] \tag{2.20}$$

If $v_r = 0$ so that the receiver is stationary, we have

$$f_r \approx f_0\left[1 + \frac{v_s}{c} + \left(\frac{v_s}{c}\right)^2\right] \tag{2.21}$$

By comparing Eq. (2.13) and Eq. (2.21), we note that the electromagnetic case and the sonic case agree only to the first power of the ratio of the source velocity to the wave velocity. We thus can see that the Doppler shift for electromagnetic and for sonic waves can be considered to be given by Eq. (2.4) only if the source and receiver velocities are small compared with the corresponding wave velocity. Although this is normally true for electromagnetic waves (see footnote 2 for some exceptions), it is not always true for sonic waves.[3]

The Doppler effect of electromagnetic waves has also been used in rheology to measure the flow of fluids. In one application a laser is used to produce an intense monochromatic light. Water flow and gas flow are determined by measuring the Doppler shift of the light scattered from particles carried in the water or the gas. In applications such as this in which the reflected wave from a transmitter is observed, the time delay is the round trip time, which is $t_0 = 2r/c$ so that instead of Eq. (2.4) we have

$$\delta = \omega - \omega_0 = -\frac{2\omega_0}{c}\frac{dr}{dt} \tag{2.22}$$

The Doppler shift obtained when observing the reflection of a transmitted wave from a moving object instead of from a moving source is observed to be twice that obtained from a moving radiating source.

The narrow laser beam affords a high spatial resolution that is used in the study of turbulent flow such as the determination of turbulence profiles required for design studies in wind tunnels. Another rheological application of importance to flying aircraft is the detection of air turbulence. Clear air turbulence is detected using Doppler laser radar operating in the near-infrared (a wavelength of about 2 μm). The reflection is obtained from dust and other aerosol particles in the air. Doppler radar is also used to detect wind shear and convective turbulence generated by thunderstorms. The airflow in various weather conditions such as tornadoes and hurricanes also is studied using Doppler radar. These few examples illustrate a few of the many and varied applications of the Doppler effect.

---

[3] Some approximate sonic wave velocities in meters per second are 345 in air, 1480 in water, from 1400 to 1600 in human tissue, and 4080 in bone.

Chapter 3

Aircraft Doppler Stabilization and Navigation

A N AIRBORNE Doppler radar is an airborne radar in which the Doppler effect is used to determine the aircraft velocity. Such a radar also can be used for the navigation and stabilization of an aircraft in flight by sending a sinusoidal wave toward the Earth's surface and using the frequency shift of the reflected wave for navigation and stabilization. For this we'll first review the basic concepts used in an airborne continuous wave (CW) Doppler radar. Then the basic method by which the navigation and the stabilization of an aircraft can be accomplished with the use of a CW Doppler radar will be discussed. The accuracy by which this can be accomplished is determined by the width of the spectrum of the reflected waveform. In order not to obscure the basic theory of how an airborne Doppler radar can be used for navigation and stabilization, we'll defer a discussion of the reflected waveform spectrum until the next chapter.

I. Basic Concepts

The first uses of radar was for detecting and determining the range of a target. In fact, this application is the source of the acronym *radar*, which derives from the first letters of the words radio detection and ranging. The target range was determined by measuring the time interval between a transmitted pulse and its echo from the target. The round trip time is

$$\tau = \frac{2r}{c} \tag{3.1}$$

in which r is the distance to the target and c is the velocity of the wave, which is approximately 3×10^8 meters/s (186,300 miles/s) for an electromagnetic wave in free space.

Early on, it was realized that the target velocity also could be determined by using the Doppler effect, which, as we have discussed, is an apparent frequency shift of a reflected wave resulting from the target velocity. However, this development had to wait for the development of oscillators that can generate sufficient power at a very high frequency with good frequency stability. The oscillator frequency stability is important because the Doppler frequency shift is not large so that the Doppler effect would be obscured by a wandering oscillator frequency.

To discuss how the Doppler effect can be used for the navigation (avigation) and stabilization of an aircraft, consider an aircraft flying at an altitude h in the x direction with a velocity v as depicted in Fig. 3.1. Let the antenna of a CW Doppler radar on the aircraft be directed at the point p in the diagram. The CW Doppler radar transmits a sinusoidal wave with the frequency ω_0. From Eq. (2.22), the Doppler shift is

$$\delta = -\frac{2\omega_0}{c}\frac{dr}{dt} \quad \text{rad/s} \tag{3.2}$$

From the diagram above, $r^2 = h^2 + r_x^2 + r_y^2$. Because h and r_y are fixed we have

$$2r\frac{dr}{dt} = 2r_x\frac{dr_x}{dt}$$

or

$$\frac{dr}{dt} = \frac{r_x}{r}\frac{dr_x}{dt} = -\frac{r_x}{r}v \tag{3.3}$$

Again from the diagram, we observe that

$$r_x = r \sin \psi_r \cos \psi_a \tag{3.4}$$

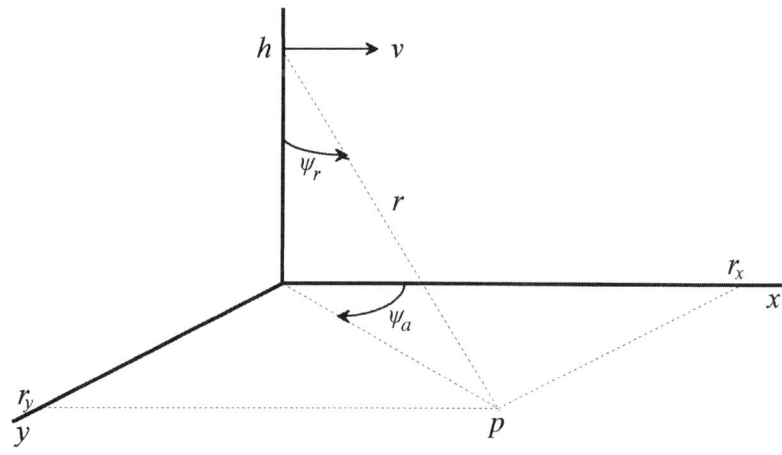

Fig. 3.1   Geometrical representation of a flying aircraft.

in which ψ_r is the range angle and ψ_a is the azimuth angle. Substituting this result in Eq. (3.3), we obtain

$$\frac{dr}{dt} = -v \sin \psi_r \cos \psi_a \tag{3.5}$$

The Doppler shift thus is

$$\delta = -\frac{2\omega_0}{c} \frac{dr}{dt} = \frac{2\omega_0}{c} v \sin \psi_r \cos \psi_a \quad \text{rad/s} \tag{3.6}$$

Because, $c = \lambda f_0$, this expression also can be written in terms of the wavelength of the radiated wave λ as

$$\delta = 2\pi \frac{2v}{\lambda} \sin \psi_r \cos \psi_a \quad \text{rad/s} \tag{3.7}$$

Because $\omega = 2\pi f$ in which f is in hertz, we also have

$$\delta = \frac{2v}{\lambda} \sin \psi_r \cos \psi_a \quad \text{hertz} \tag{3.8}$$

The aircraft navigation and control will be obtained by determining the Doppler shift δ at various range and elevation angles. Unfortunately, even though the transmitted wave is a sinusoid, the echo will not be a sinusoid with a frequency that is the transmitted frequency ω_0 shifted by δ. Rather, the echo will be a nonsinusoidal waveform. The reason is that the echo is the sum of the reflections from many terrain particles in the antenna beam. Observe that dr/dt varies as a particle travels through the antenna beam and it also differs from particle to particle. In addition, the return from a given particle is amplitude modulated by the antenna pattern and with the manner in which the backscattering from the particle varies with angle. We thus can view the echo as a complex waveform that is the sum of many sinusoids, each of which has been amplitude and frequency modulated. We would like the spectrum width to be narrow and centered at the frequency $\omega_0 + \delta$ so that the echo can be approximated as a sinusoid with the frequency $\omega_0 + \delta$. The exact echo spectrum will be determined later in this text where we shall determine its center frequency and width.

II. Aircraft Navigation and Stabilization

In order not to obscure our discussion of the manner by which an aircraft can be stabilized and navigated using an airborne Doppler radar, we shall simplify our discussion by considering the echo to be a sinusoid with a frequency that has been shifted from the transmitted frequency by the Doppler shift given by Eq. (3.8). In practical application, the actual spectrum width causes a small random error in the measurements that can be determined using the exact theory developed later in this text. The exact theory enables one to minimize this error by choosing parameter values to maximize the ratio of the Doppler shift to the spectrum bandwidth.

The following is a possible procedure for aircraft stabilization solely from Doppler measurements. The stabilization and navigation of an aircraft requires four parameters, roll, pitch, drift, and velocity, to be determined. The roll is the angle of rotation ϕ_r of the aircraft about the fuselage; facing forward, a positive angle is a counterclockwise rotation. The pitch is the angle ϕ_p at which the aircraft is nosed up or down about its transverse axis; a positive angle is a nose-up angle. The drift is the angular modification of the aircraft course direction due to wind currents. The aircraft heading is the direction relative to the ground of the aircraft axis. Because of the wind currents, the actual velocity differs from the aircraft heading by the drift angle ϕ_d. Figure 3.2 depicts the case for which the drift angle is positive. The velocity v is the aircraft ground speed, which is the aircraft velocity relative to the ground.

Because there are four quantities to determine, we require at least four Doppler measurements. For this, the airborne Doppler radar is designed with four antennae. To begin, assume the aircraft roll, pitch, and drift are zero. Then, referring to Fig. 3.1, let one antenna beam be directed toward the terrain fore of the aircraft with the angles $\psi_r = \theta_e$ and $\psi_a = 0$; let a second antenna beam be directed toward the terrain aft of the aircraft with the angles $\psi_r = -\theta_e$ and $\psi_a = 0$; let a third antenna beam be directed toward the terrain on the starboard side of the aircraft with the angles $\psi_r = \theta_a$ and $\psi_a = \pi/2$; and let a fourth antenna beam be directed toward the terrain on the port side of the aircraft with the angles $\psi_r = \theta_a$ and $\psi_a = -\pi/2$. With this configuration, note that aircraft roll does not affect the fore and aft beams and the aircraft pitch does not affect the port and starboard antenna beams.

Now, still considering the drift angle to be zero, the angles for each of the beams due to the aircraft roll and pitch are:

1) For the fore antenna beam, $\psi_r = \theta_e + \phi_p$ and $\psi_a = 0$
2) For the aft antenna beam, $\psi_r = -\theta_e + \phi_p$ and $\psi_a = 0$
3) For the starboard beam, $\psi_r = \theta_a + \phi_r$ and $\psi_a = \pi/2$
4) For the port beam, $\psi_r = \theta_a - \phi_r$ and $\psi_a = -\pi/2$

Observe that the drift just constitutes a rotation of the coordinates, which does not affect the range angle ψ_r but increases the azimuth angle ψ_a by ϕ_d. With the aircraft drift included, the range and azimuth angles then are:

1) For the fore antenna beam, $\psi_r = \theta_e + \phi_p$ and $\psi_a = \phi_d$
2) For the aft antenna beam, $\psi_r = -\theta_e + \phi_p$ and $\psi_a = \phi_d$
3) For the starboard beam, $\psi_r = \theta_a + \phi_r$ and $\psi_a = \pi/2 + \phi_d$
4) For the port beam, $\psi_r = \theta_a - \phi_r$ and $\psi_a = -\pi/2 + \phi_d$

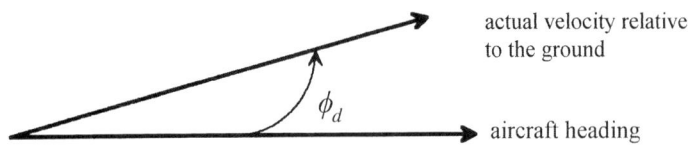

Fig. 3.2   Illustrating a positive drift angle.

The Doppler shift associated with each of the four antenna beams now can be obtained from Eq. 3.8. The Doppler shifts in hertz thus are:

1) For the fore antenna beam,

$$\delta_f = \frac{2v}{\lambda} \sin(\theta_e + \phi_p) \cos(\phi_d) \quad \text{hertz} \qquad (3.9a)$$

2) For the aft antenna beam,

$$\delta_a = \frac{2v}{\lambda} \sin(-\theta_e + \phi_p) \cos(\phi_d)$$

$$= -\frac{2v}{\lambda} \sin(\theta_e - \phi_p) \cos(\phi_d) \quad \text{hertz} \qquad (3.9b)$$

3) For the starboard antenna beam,

$$\delta_S = \frac{2v}{\lambda} \sin(\theta_a + \phi_r) \cos\left(\frac{\pi}{2} + \phi_d\right)$$

$$= -\frac{2v}{\lambda} \sin(\theta_a + \phi_r) \sin(\phi_d) \quad \text{hertz} \qquad (3.9c)$$

4) For the port antenna beam,

$$\delta_p = \frac{2v}{\lambda} \sin(\theta_a - \phi_r) \cos\left(-\frac{\pi}{2} + \phi_d\right)$$

$$= \frac{2v}{\lambda} \sin(\theta_a - \phi_r) \sin(\phi_d) \quad \text{hertz} \qquad (3.9d)$$

Using standard trigonometric identities we then obtain

$$\delta_f + \delta_a = \frac{4v}{\lambda} \cos\phi_d \sin\phi_p \cos\theta_e \qquad (3.10a)$$

$$\delta_f - \delta_a = \frac{4v}{\lambda} \cos\phi_d \cos\phi_p \sin\theta_e \qquad (3.10b)$$

$$\delta_p + \delta_S = -\frac{4v}{\lambda} \sin\phi_d \sin\phi_r \cos\theta_a \qquad (3.10c)$$

$$\delta_p - \delta_S = \frac{4v}{\lambda} \sin\phi_d \cos\phi_r \sin\theta_a \qquad (3.10d)$$

We then obtain by dividing one equation by another

$$\tan\phi_p = \frac{\delta_f + \delta_a}{\delta_f - \delta_a} \tan\theta_e \qquad (3.11a)$$

$$\tan \phi_r = -\frac{\delta_p + \delta_s}{\delta_p - \delta_s} \tan \theta_a \qquad (3.11b)$$

$$\tan \phi_d = -\frac{\delta_p + \delta_s}{\delta_f + \delta_a} \frac{\cos \theta_e \sin \phi_p}{\cos \theta_a \sin \phi_r} \qquad (3.11c)$$

We now can determine the pitch angle ϕ_p from Eq. (3.11a) and the roll angle ϕ_r from Eq. (3.11b). Once these two angles are determined, the drift angle ϕ_d can be determined from Eq. (3.11c). The ground velocity v then can be determined from any one of Eqs. (3.9). To reduce the algebraic manipulations, the port and starboard beams were chosen to be at right angles to the aircraft. However, note that if the drift angle ϕ_d is small, then there could be unacceptable error because then δ_p and δ_s will be small. In practice, it thus may be desirable to choose $|\psi_a| < \pi/2$ radians.

Note that the plane can be kept on level flight by using the control described above to hold the pitch angle at zero degrees. Observe that a rise of the terrain is equivalent to a nose down of the aircraft and a fall of the terrain is equivalent to a nose up of the aircraft. We thus observe that using the control to keep the aircraft on level flight also will tend to fly the aircraft parallel the terrain.

All of these calculations can be performed by a simple microprocessor whose output then can be used in a feedback arrangement to stabilize the aircraft. The microprocessor output also can be used to navigate the aircraft. The errors associated with this Doppler stabilization and navigation system depend on the accuracy with which the four Doppler shifts are measured. As I mentioned above, this depends on the width of the Doppler spectrum and whether the center frequency of the Doppler spectrum is at the transmitted frequency plus the appropriate Doppler shift. Clearly, the greater the ratio of the Doppler frequency to the spectrum width, the smaller will be the error. All these quantities will be determined later in this text. The optimum values of θ_e and θ_a for which the errors are a minimum can be determined from those results.

Chapter 4

Quasi-Static Approximation of the Doppler Spectrum

I. Introduction

IN ORDER to clearly discuss how Doppler radar can be used for aircraft navigation and stabilization, the echo obtained by an aircraft Doppler radar in Chapter 3 was considered to be a sinusoid with a constant amplitude and frequency. The frequency of the echo was considered to be equal to that of the sinusoid transmitted by the Doppler radar transmitter displaced by an amount equal to the Doppler shift that would be obtained from an individual scatterer when it is at the center of the antenna beam. The spectrum was thus assumed to contain only one frequency in order not to obscure the basic theory of how an airborne Doppler radar can be used for navigation and stabilization.

As we discussed, the terrain is actually composed of many randomly distributed individual scatterers. As a given scatterer passes through the radar antenna beam, the reflection is actually a sinusoid that has been phase and amplitude modulated. The phase modulation is the result of the changing path length between the radar and the scatterer and the changing phase of the reflection from the scatterer as it passes through the radar antenna beam.[1] The amplitude modulation is a consequence of the scatterer passing through the radar antenna pattern and the changing scatterer reflectivity. The echo received by the radar from the ground is the sum of the reflections received from all the randomly distributed scatterers. Consequently, the spectrum of the received waveform is not concentrated at a single frequency. An airborne Doppler radar is thus designed to track the center frequency of the received waveform spectrum. The difference between this center frequency and the frequency of the sinusoid radiated by the radar is called the "Doppler shift." The accuracy of the tracking of this Doppler shift is determined by the width of the spectrum because the larger the spectrum width, the less accurate can the spectrum center frequency be determined. For this reason, it is important to know the exact relation of the spectrum center frequency to the aircraft ground velocity and to know the width of the spectrum in order to

---

[1] I assume throughout this text that the terrain scatterers are not moving. This occurs with wind-blown vegetation such as trees and plants. Motion would contribute to random phase fluctuation, which would affect the Doppler bandwidth slightly.

be able to determine the accuracy with which the center frequency can be tracked. The two spectrum parameters of major importance are thus the spectrum center frequency δ and the spectrum bandwidth σ.

Classically, a quasi-static approximation was used to determine the Doppler spectrum. I call the approximation a quasi-static one because the approximation is that the rate of change of the amplitude and frequency of the reflection from a scatterer is sufficiently small so that it can be ignored. This approximation is similar to the classical approximation of the spectrum of a frequency-modulated sinusoid. If the sinusoidal frequency is changed very slowly, then a good approximation of the power density spectrum is a quasi-static one. That is, the fraction of the average power in a given frequency interval is proportional to the fraction of time the sinusoidal frequency is in the given interval. Of course, the more rapidly the sinusoidal frequency is changed, the poorer is this approximation and so, for example, it is an unacceptable approximation for the determination of the spectrum of an FM radio transmission. However, this quasi-static approximation for the spectrum of a frequency-modulated sinusoid is so intuitive and simple to determine that it is a desirable approximation for those situations for which it is valid; unfortunately, useful conditions for its validity have not been determined.

For similar reasons, the quasi-static approximation for the aircraft Doppler spectrum is desirable. However, without an exact theory, the conditions for which the quasi-static approximation is valid are not known. To use the quasi-static approximation with any confidence, the approximation error must be known. Because the determination of the quasi-static approximation is rather straightforward, it will be presented in this chapter. The exact theory will be developed in a later chapter after a discussion of some tools needed for its development. From physical arguments, we will determine conditions for which the quasi-static approximation error should be small. In this manner, the quasi-static approximation will serve as a check of the exact theory. This is similar to the manner by which Einstein checked his special relativity theory by arguing that his theory should correspond to the Newton theory for small velocities. This technique is generally a good one to use in checking a new theory.

II.   Quasi-static Approximation

From the above discussion, the quasi-static approximation of the Doppler spectrum is to assign to each point on the terrain a Doppler frequency that corresponds to the rate of change of the radial distance to that point and to calculate the spectrum in a given frequency interval by summing the total power received from all infinitesimal regions with an assigned frequency in that interval. Note that this assumes that reflections from the various scatterers are uncorrelated so that the total power of the echo received by the Doppler radar is equal to the sum of the powers received from the individual scatterers. Also, the spectrum of the power received from a scatterer is the quasi-static approximation discussed above.

We now must determine the Doppler frequency corresponding to each point P on the terrain. As depicted in Fig. 4.1, the aircraft is above the origin of the $x-y$ plane and traveling in the x direction with a velocity v. Its altitude above the average terrain level is h. The center of the antenna beam on the $x-y$ plane

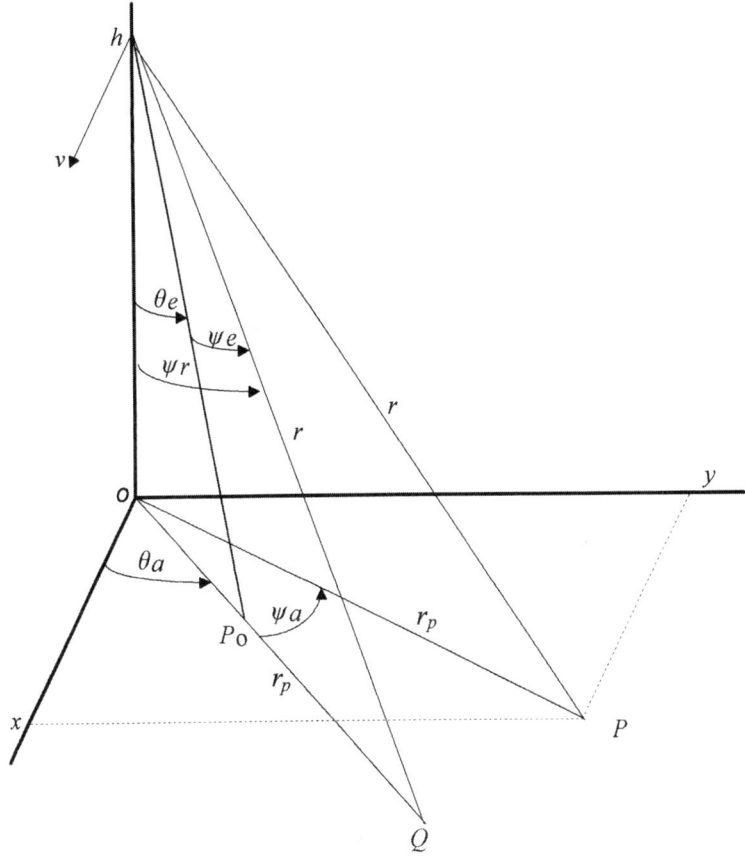

Fig. 4.1   Geometrical representation of a flying aircraft.

is at the point P_0. The line from the antenna at $(0, 0, h)$ to the point P_0, which is at $(x_0, y_0, 0)$, thus corresponds to the antenna beam axis. The antenna is gimbaled so that it can be rotated in azimuth and elevation. The antenna shown in Fig. 4.1 has been rotated θ_a radians in azimuth and θ_e radians in elevation. Also, ψ_r is the range angle and ψ_a is the azimuth angle from the antenna to a point on the terrain.

We first determine the Doppler frequency to assign to each infinitesimal region of the terrain in terms of angles from the airborne Doppler radar antenna. From Eq. (3.2) and Eq. (3.3) and using the relation $\lambda f_0 = c$, in which λ is the wavelength of the radiated sinusoid, we obtain

$$\delta = \omega - \omega_0 = -\frac{2\omega_0}{c}\frac{dr}{dt} = -\frac{4\pi}{\lambda}\frac{dr}{dt} = -\frac{4\pi}{\lambda}\frac{\partial r}{\partial x}\frac{dx}{dt} = -\frac{4\pi x}{\lambda}\frac{dx}{r\,dt} = \frac{4\pi v}{\lambda}\frac{x}{r} \quad (4.1)$$

We now must determine the Doppler frequency corresponding to each point P on the terrain. Consider a point P as shown in Fig. 4.1 within the antenna beam on the x–y plane that is at a distance r from the antenna. We then obtain the following relations from the right triangle h-o-Q with the hypotenuse of length r.

$$r = h \sec \psi_r \tag{4.2a}$$

and

$$r_p = r \sec \psi_r \tag{4.2b}$$

Also from Fig. 4.1,

$$x = r_p \cos (\theta_a + \psi_a) \tag{4.3a}$$

and

$$y = r_p \sin (\theta_a + \psi_a) \tag{4.3b}$$

Substituting, we obtain the following relations

$$r = h \sec \psi_r \tag{4.4a}$$
$$x = r \sin \psi_r \cos (\theta_a + \psi_a) \tag{4.4b}$$

and

$$y = r \sin \psi_r \sin (\theta_a + \psi_a) \tag{4.4c}$$

Substituting Eq. (4.4), in Eq. (4.1), we obtain the Doppler frequency, $\delta(\psi_r, \psi_a)$ to assign an infinitesimal region at an angle (ψ_r, ψ_a)

$$\delta(\psi_r, \psi_a) = \frac{4\pi v}{\lambda} \sin \psi_r \cos (\theta_a + \psi_a) \text{ rad/s} \tag{4.5}$$

Note that the Doppler frequency assigned to the beam center for which $\psi_a = 0$ and $\psi_r = \theta_e$ is

$$\delta(\theta_e, 0) = \frac{4\pi v}{\lambda} \sin \theta_e \cos \theta_a \text{ rad/s} \tag{4.6}$$

which is our previous result, Eq. (3.7).

Now, the effective two-way antenna pattern is

$$K(\psi_r, \psi_a) = S(\psi_r, \psi_a) B(\psi_r, \psi_a) \tag{4.7}$$

in which $S(\psi_r, \psi_a)$ is the actual two-way antenna pattern and $B(\psi_r, \psi_a)$ is the expected backscatter from the terrain.

The quasi-static approximation of the Doppler power density spectrum $P(\omega)$ as a function of ψ_r and ψ_e is

$$P(\psi_r, \psi_a) = \frac{1}{r^4} K^2(\psi_r, \psi_a) \qquad (4.8)$$

The factor r^4 accounts for the wave amplitude decrease as it travels the distance r to the point P on the terrain and then, after reflection, travels the same distance r back to the antenna.

The Doppler frequency is the mean frequency of the Doppler spectrum. Its quasi-static approximation, δ_{qs}, is the first moment of the quasi-static Doppler spectrum. That is

$$\delta_{qs} = \frac{\int_0^\infty \omega P(\omega)d\omega}{\int_0^\infty P(\omega)d\omega} \text{ rad/s} \qquad (4.9)$$

To express this equation in terms of angles, we have from Eq. (4.8) that the total power of the echo received by the Doppler radar is

$$\int_0^\infty P(\omega)d\omega = \int_{-\pi/2}^{\pi/2} \int_{-\pi/2}^{\pi/2} P(\psi_r, \psi_a)d\psi_r\, d\psi_a$$
$$= \int_{-\pi/2}^{\pi/2} \int_{-\pi/2}^{\pi/2} \frac{1}{r^4} K^2(\psi_r, \psi_a)d\psi_r\, d\psi_a \qquad (4.10)$$

The numerator in Eq. (4.9) also can be expressed in terms of the angles from the radar antenna as

$$\int_0^\infty \omega P(\omega)d\omega = \int_{-\pi/2}^{\pi/2} \int_{-\pi/2}^{\pi/2} P(\psi_r, \psi_a)\delta(\psi_r, \psi_a)d\psi_r\, d\psi_a$$
$$= \int_{-\pi/2}^{\pi/2} \int_{-\pi/2}^{\pi/2} \frac{1}{r^4} K^2(\psi_r, \psi_a)\delta(\psi_r, \psi_a)d\psi_r\, d\psi_a \qquad (4.11)$$

Equation (4.9) thus can be expressed in the form

$$\delta_{qs} = \frac{\displaystyle\int_{-\pi/2}^{\pi/2} \int_{-\pi/2}^{\pi/2} P(\psi_r, \psi_a)\delta(\psi_r, \psi_a)d\psi_r\, d\psi_a}{\displaystyle\int_{-\pi/2}^{\pi/2} \int_{-\pi/2}^{\pi/2} P(\psi_r, \psi_a)d\psi_r\, d\psi_a}$$

$$= \frac{\displaystyle\int_{-\pi/2}^{\pi/2} \int_{-\pi/2}^{\pi/2} \frac{1}{r^4} K^2(\psi_r, \psi_a)\delta(\psi_r, \psi_a)d\psi_r\, d\psi_a}{\displaystyle\int_{-\pi/2}^{\pi/2} \int_{-\pi/2}^{\pi/2} \frac{1}{r^4} K^2(\psi_r, \psi_a)d\psi_r\, d\psi_a} \text{ rad/s} \qquad (4.12)$$

The bandwidth is the second moment of the Doppler spectrum about the Doppler frequency.[2] Its quasi-static approximation σ_{qs} is

$$\sigma_{qs}^2 = \frac{\int_0^\infty (\omega - \delta_{qs})^2 P(\omega)d\omega}{\int_0^\infty P(\omega)d\omega} \ (\text{rad/s})^2 \tag{4.13}$$

By expanding the numerator bracket and with the use of Eq. (4.9), we obtain from Eq. (4.13) that

$$\sigma_{qs}^2 + \delta_{qs}^2 = \frac{\int_0^\infty \omega^2 P(\omega)d\omega}{\int_0^\infty P(\omega)d\omega} \ (\text{rad/s})^2 \tag{4.14}$$

With the use of the equations developed above, this equation also can be expressed in terms of the angles from the radar antenna as

$$\sigma_{qs}^2 + \delta_{qs}^2 = \frac{\displaystyle\int_{-\pi/2}^{\pi/2} \int_{-\pi/2}^{\pi/2} \frac{1}{r^4} K^2(\psi_r, \psi_a)\delta_{qs}^2(\psi_r, \psi_a)d\psi_r\,d\psi_a}{\displaystyle\int_{-\pi/2}^{\pi/2} \int_{-\pi/2}^{\pi/2} \frac{1}{r^4} K^2(\psi_r, \psi_a)d\psi_r\,d\psi_a} \ (\text{rad/s})^2 \tag{4.15}$$

Eq. (4.12) and Eq. (4.15) are the general expressions for the quasi-static Doppler frequency and spectrum bandwidth. Note from Eq. (4.5) that $\delta(\psi_r, \psi_a)$ varies directly as v/λ because it is equal to that factor times a function of geometry only. Each of the quantities δ_{qs} and σ_{qs} in Eq. (4.12) and Eq. (4.15) thus will be equal to v/λ times a function of geometry. We thus shall use dimensionless values for our illustrations by graphing $(\lambda/v)\delta_{qs}$ and $(\lambda/v)\sigma_{qs}$. The graphs thus will be valid for any velocity and wavelength. For convenience, the graphical values for δ_{qs} and σ_{qs} will be in hertz (cycles/s).

III.   Numerical Illustration

The following numerical results are presented to help better explain the quasi-static approximation. This approximation will be compared with my exact theory in Chapter 9 in order to determine the conditions for which the quasi-static approximation is not valid and the reasons for its invalidity.

Before presenting the numerical illustrations, note that r in Eq. (4.4a) is proportional to the height h. Thus the quasi-static Doppler frequency δ_{qs} and the quasi-static Doppler bandwidth σ_{qs} do not vary with h because the same factor, r^4, is in the numerator and the denominator of Eq. (4.12) and Eq. (4.15). However, we later shall see from the exact theory that the Doppler bandwidth actually does vary with height.

---

[2] There are many definitions of bandwidth. This definition of bandwidth is a measure of the dispersion of the Doppler spectrum about the Doppler frequency.

We'll present illustrative graphs of δ_{qs} and σ_{qs} versus various antenna parameters for the case in which the terrain is a diffuse surface. A diffuse surface is one for which the backscatter follows the Lambert law, which is $b \sin \beta$ in which β is the angle between the incident wave and the plane of the terrain. A surface for which this is a good model is a smooth sandy ocean beach. In Fig. 4.1, β is the angle between the $x-y$ plane and the line from the antenna to the point P. Thus, for a terrain that is a diffuse surface, the expected terrain backscatter is[3]

$$B(\psi_r, \psi_a) = b \sin \beta = b\frac{h}{r} \qquad (4.16)$$

Substituting Eq. (4.4a) we have that the expected backscattering from a diffuse terrain in terms of the antenna angles is

$$B(\psi_r, \psi_a) = b \cos \psi_r \qquad (4.17)$$

For our illustration, we'll use a normal function for the antenna pattern, which is

$$S(\psi_r, \psi_a) = e^{(-\mu_a \psi_e^2 + 2\mu_{ae}\psi_a\psi_e + \mu_e\psi_a^2)/2\sigma_1^2\sigma_2^2} \qquad (4.18)$$

in which

$$\mu_a = \sigma_1^2 \sin^2\phi + \sigma_2^2 \cos^2\phi \qquad (4.19a)$$

$$\mu_e = \sigma_1^2 \cos^2\phi + \sigma_2^2 \sin^2\phi \qquad (4.19b)$$

$$\mu_{ae} = \left(\sigma_2^2 - \sigma_1^2\right) \sin\phi\cos\phi = \frac{1}{2}\left(\sigma_2^2 - \sigma_1^2\right) \sin 2\phi \qquad (4.19c)$$

and, as shown in Fig. 4.1,

$$\psi_e = \psi_r - \theta_e \qquad (4.20)$$

This antenna pattern was chosen for our illustration because it is a good approximation of many actual antenna patterns in use. The one we have chosen for our illustration does not contain side lobes. Side lobes would have to be incorporated into the pattern if it is desired to study their effect on the Doppler power density spectrum.

The cross section of our antenna pattern, which is the level curve of $S(\psi_r, \psi_a)$ in the (ψ_r, ψ_a) plane, are ellipses centered at $(\psi_r, \psi_a) = (\psi_e, 0)$ as shown in Fig. 4.2. As shown, one elliptical axis is at an angle ϕ from the ψ_e axis and

---

[3] Note that the Lambert law simply assumes that the backscatter is proportional to the projected area.

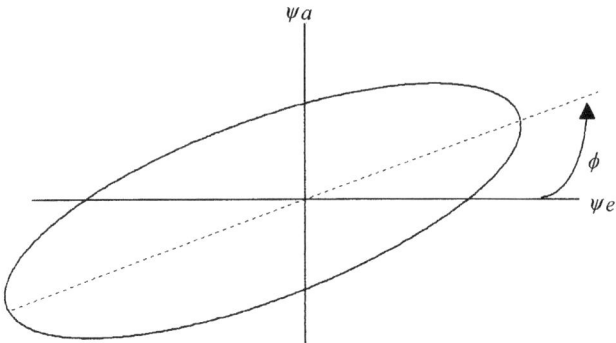

Fig. 4.2   Level curve of the antenna pattern cross section.

has a length proportional to σ_1. The length of the other elliptical axis is proportional to σ_2.

These equations have been programmed using MatLab 6.3. Before running the program, you should read Chapter 10, which includes a description of the computer program and its use; also, a copy of the program is included at the end of this text. For computational efficiency, C++ is used for the numerical computations; the numerical results are then used by MatLab to produce the graphs. We'll determine the effect of various parameters on the Doppler power density spectrum using the quasi-static approximation. In addition to the determination using the quasi-static approximation, the computer program also allows the determination using my exact theory. We'll only use the quasi-static approximation now. Later, after I present my exact theory, we'll determine the error incurred using the quasi-static approximation error by comparing these results with those obtained with the exact theory.

The quasi-static approximation of the Doppler frequency δ_{qs}, given by Eq. (4.12), is the mean frequency of the quasi-static Doppler spectrum. Thus, for small antenna beamwidths, we would expect that the frequency assigned to the beam center given by Eq. (4.6), $\delta(\theta_e, 0)$, is a good approximation to δ_{qs}.

The program plots of the quasi-static Doppler frequency are graphs of the normalized quasi-static Doppler frequency, which is $(\lambda/v)\delta_{qs}$. The reason for this normalization is that the graphs are then independent of wavelength and aircraft velocity. Furthermore, for convenience in using the graphs, δ_{qs} is in hertz. From Eq. (4.6) the normalized frequency assigned to the beam center thus is $2\sin\theta_e\cos\theta_a$ cycles.

Figure 4.3 is a graph of the normalized quasi-static Doppler frequency versus the elevation angle θ_e for the azimuth angle $\theta_a = 0, 25, 50,$ and 75 degrees. Figure 4.4 is a graph of the normalized quasi-static Doppler frequency versus the azimuth angle θ_a for the elevation angle $\theta_e = 5, 30, 55,$ and 80 degrees. For both graphs, the antenna pattern has a circular cross section with $\sigma_1 = \sigma_2 = 1$ degree. Observe that indeed $\delta_{qs} \approx \delta(\theta_e, 0)$. It can be observed from plots with different values of σ that this approximation is valid for beamwidths σ less than 3 degrees.

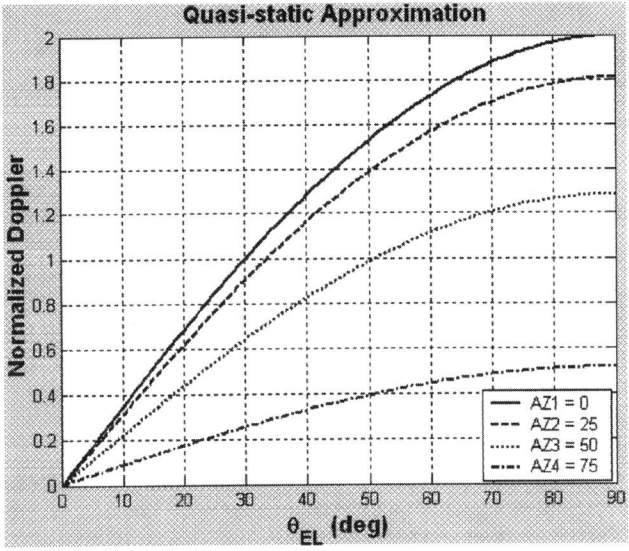

Fig. 4.3 Normalized quasi-static Doppler frequency versus the elevation angle.

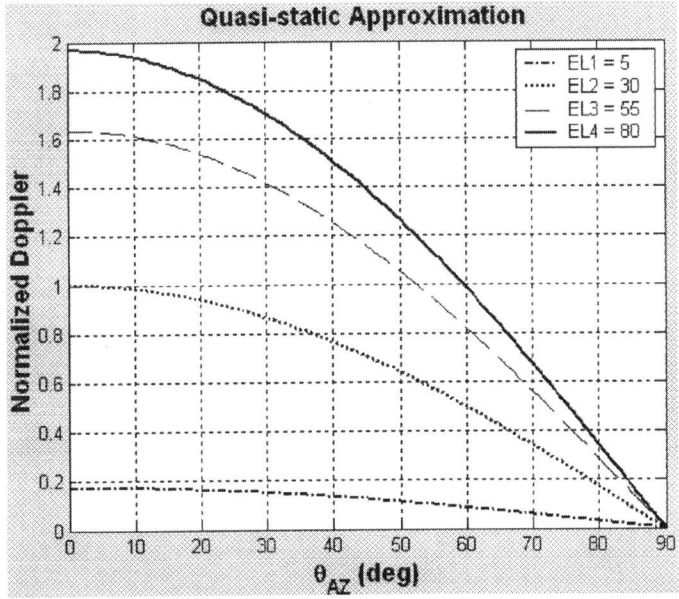

Fig. 4.4 Normalized quasi-static Doppler frequency versus the azimuth angle.

Note that σ is not the 3-dB beamwidth. Because $S(\psi_r, \psi_a)$ is a two-way antenna pattern, the 3-dB beamwidth is the angle at which $S(\psi_r, \psi_a) = 1/2$. Now $S(\psi_r, \psi_a)$ in Eq. (4.18) is a normal function so that for $\phi = 0$, the 3-dB width along the σ_1 axis is $\psi_e = \sigma_1 \sqrt{2 \ln 2} = 1.17741\sigma_1$ and is $\psi_a = \sigma_2 \sqrt{2 \ln 2} = 1.7741\sigma_2$ along the σ_2 axis. Thus, for example, a beamwidth with $\sigma = 3$ degrees corresponds to 3-dB beamwidth of 3.53 degrees.

Figure 4.5 is a graph of the normalized quasi-static Doppler frequency versus σ_1 for $\sigma_2 = 0.1, 4.1$, and 8.1 degrees with the rotation angle $\phi = 0$, the azimuth angle $\theta_a = 60$ degrees, and the elevation angle $\theta_e = 30$ degrees. As expected, δ_{qs} is not a strong function of the beamwidth. Also, as discussed previously, δ_{qs} does not vary with height. Figure 4.6 is a graph of the normalized quasi-static Doppler frequency versus the rotation angle ϕ of the antenna ellipse for $\sigma_1 = 2$ degrees and $\sigma_2 = 1$ degree with the azimuth angle $\theta_a = 60$ degrees and the elevation angle $\theta_e = 30$ degrees. Note that the quasi-static Doppler frequency is not a strong function of ϕ as we should expect because it is not a strong function of the beamwidth.

We now examine the quasi-static bandwidth σ_{qs}, which is a measure of variation of $\delta(\psi_r, \psi_a)$ given by Eq. (4.5) over the surface illuminated by the antenna. As with the graphs of the quasi-static Doppler frequency, the program graphs of the quasi-static Doppler bandwidth are graphs of the normalized quasi-static Doppler bandwidth, which is $(\lambda/v)\sigma_{qs}$. Again, the reason for this normalization is that the graphs are then independent of wavelength and aircraft velocity. Furthermore, for convenience in using the graphs, σ_{qs} is in hertz. Note that the shape of the quasi-static Doppler spectrum is not necessarily that of a normal function such as that of the antenna pattern we are using. The 3-dB bandwidth will be proportional to σ_{qs} but, as will be seen from our discussion, its exact

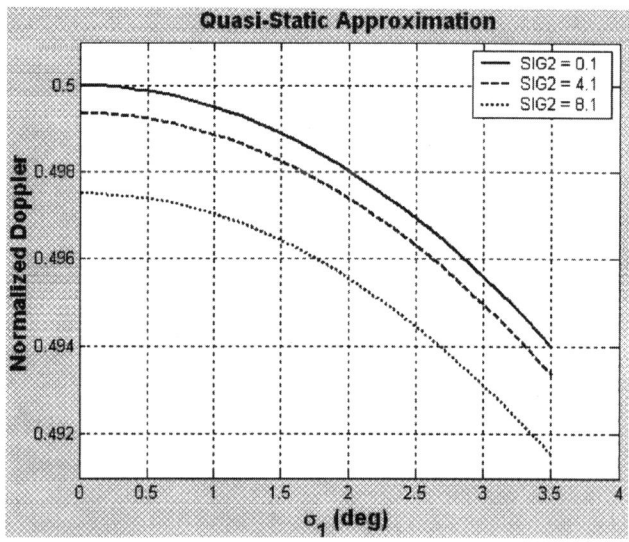

Fig. 4.5 Normalized quasi-static Doppler frequency versus σ_1.

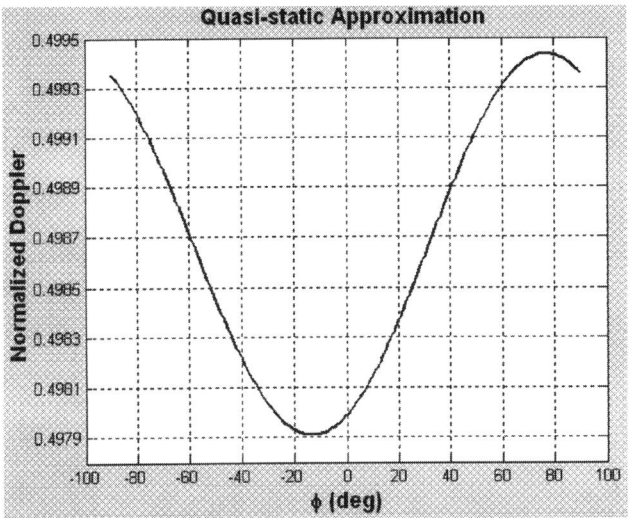

Fig. 4.6   Normalized quasi-static Doppler frequency versus the rotation angle, ϕ.

value is not really needed. However, the value of the proportionality factor can be calculated from the shape of the quasi-static Doppler spectrum, which can be determined from Eq. (4.8).

First, as discussed, σ_{qs} and δ_{qs} do not vary with height. Figure 4.7 is a graph of the normalized quasi-static Doppler bandwidth versus the elevation angle θ_e for the

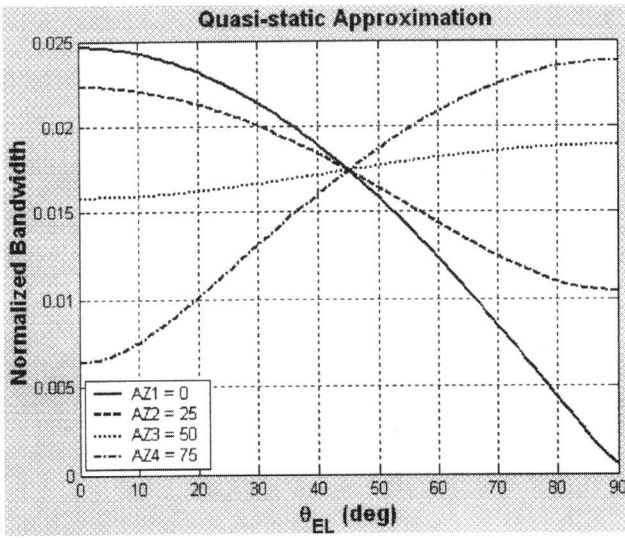

Fig. 4.7   Normalized quasi-static Doppler bandwidth versus the elevation angle.

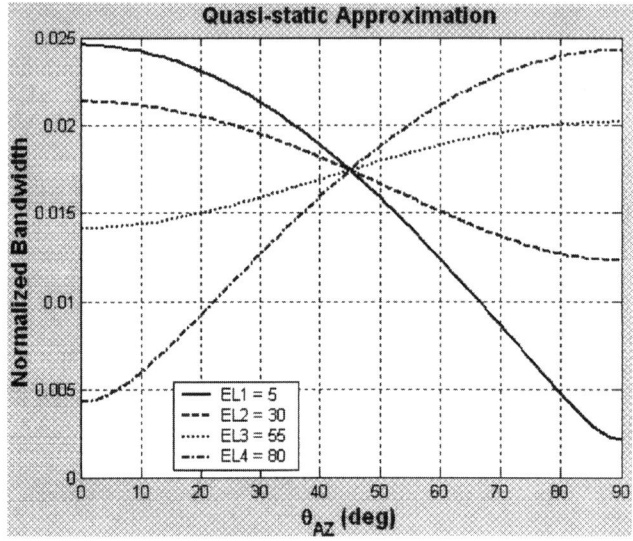

Fig. 4.8   Normalized quasi-static Doppler bandwidth versus the azimuth angle.

azimuth angle $\theta_a = 0$, 25, 50, and 75 degrees. Figure 4.8 is a graph of the normalized quasi-static Doppler bandwidth versus the azimuth angle θ_a for the elevation angle $\theta_e = 5$, 30, 55, and 80 degrees. For both graphs, the antenna pattern has a circular cross section with $\sigma_1 = \sigma_2 = 1$ degree. To understand these graphs, note from Fig. 4.3 and Fig. 4.4 that δ_{qs} decreases for increasing θ_a and the rate of increase decreases for increasing θ_e; also δ_{qs} increases for increasing θ_e and the rate of increase increases for increasing θ_a. These two effects compete to determine the variation of $\delta(\psi_r, \psi_a)$ over the region illuminated by the antenna to result in the graphs shown. Note then that the graphs of the quasi-static bandwidth versus θ_e or θ_a could be significantly different for $\sigma_1 \neq \sigma_2$ and for $\phi \neq 0$.

Figure 4.9 is a graph of the normalized quasi-static Doppler bandwidth versus σ_1 for $\sigma_2 = 0.1$, 1.1, 2.1, and 3.1 degrees with the rotation angle $\phi = 0$, the azimuth angle $\theta_a = 60$ degrees, and the elevation angle $\theta_e = 30$ degrees. As expected, the bandwidth increases with increasing σ_1 because the larger the beamwidth, the larger the variation of $\delta(\psi_r, \psi_a)$ over the region illuminated by the antenna. Note that as σ_1 decreases, the bandwidth becomes asymptotic to a value determined by σ_2; the smaller σ_2, the smaller the asymptotic value. Also observe that the bandwidth increases linearly with σ_1 for $\sigma_1 > \sigma_2$. Figure 4.10 is a graph of the normalized quasi-static Doppler bandwidth versus σ_2 for $\sigma_1 = 0.1$, 1.1, 2.1, and 3.1 degrees with the rotation angle $\phi = 0$, the azimuth angle $\theta_a = 60$ degrees, and the elevation angle $\theta_e = 30$ degrees. With σ_1 and σ_2 interchanged, note that this graph is identical with that of Fig. 4.9.

Because the quasi-static Doppler bandwidth is a measure of the variation of $\delta(\psi_r, \psi_a)$ over the region illuminated by the antenna, there should be an

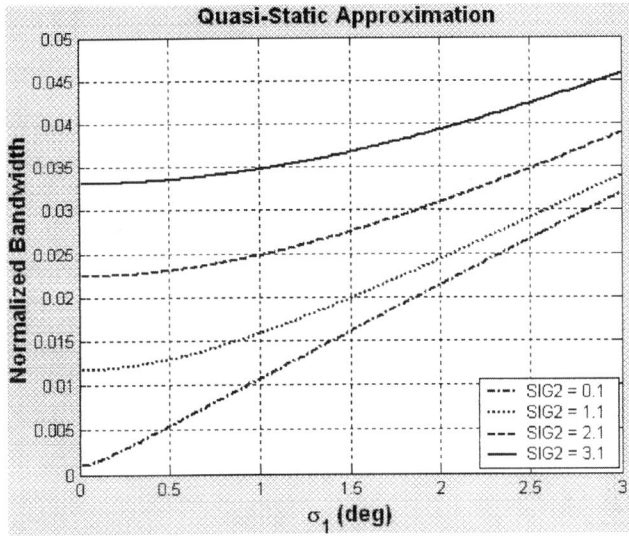

Fig. 4.9 Normalized quasi-static Doppler bandwidth versus σ_1.

orientation of the antenna pattern ellipse for which the bandwidth is a minimum. This minimum should be at an angle at which the major elliptical axis lies tangent to an isodoppler. An isodoppler is a curve along which the Doppler $\delta(\psi_r, \psi_a)$ is a constant.

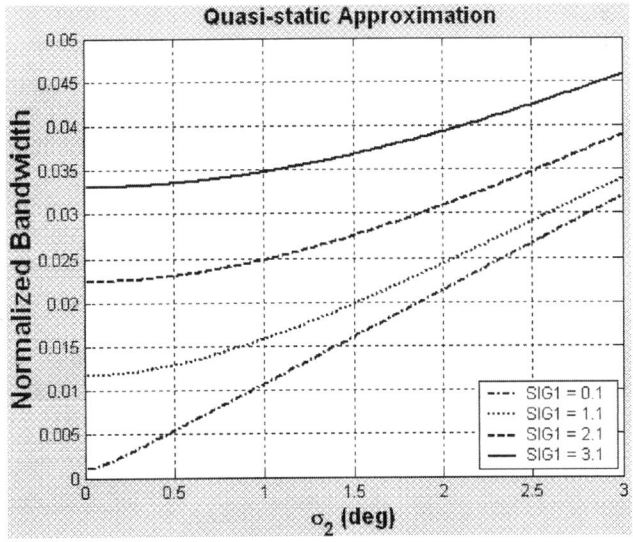

Fig. 4.10 Normalized quasi-static Doppler bandwidth versus σ_2.

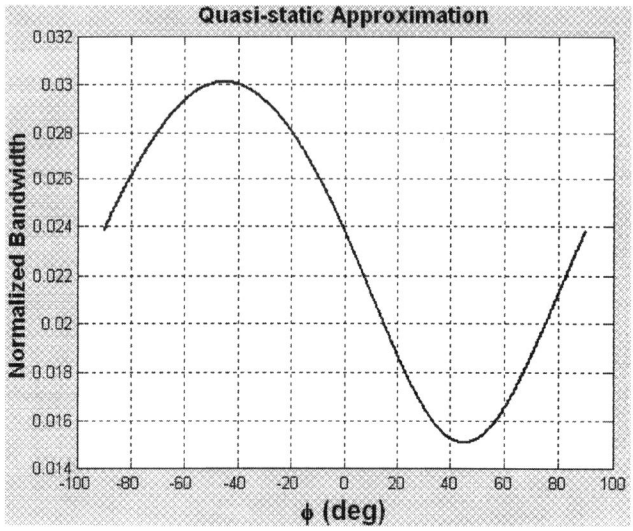

Fig. 4.11   Normalized quasi-static Doppler bandwidth versus the rotation angle, ϕ.

To determine the tangent to the isodoppler, we have from Eq. (4.5) and Eq. (4.20) that

$$\delta(\psi_r, \psi_a) = \frac{2v}{\lambda} \sin(\theta_e + \psi_e) \cos(\theta_a + \psi_a) \text{ hertz} \qquad (4.21)$$

The change in δ, $d\delta$, due to a small change in ψ_e and ψ_a is

$$d\delta = \frac{2v}{\lambda} [\cos(\theta_e + \psi_e) \cos(\theta_a + \psi_a) d\psi_e$$
$$- \sin(\theta_e + \psi_e) \sin(\theta_a + \psi_a) d\psi_a] \qquad (4.22)$$

Along an isodoppler $d\delta = 0$, so we require

$$\cos(\theta_e + \psi_e) \cos(\theta_a + \psi_a) d\psi_e = \sin(\theta_e + \psi_e) \sin(\theta_a + \psi_a) d\psi_a \qquad (4.23)$$

The slope of the tangent, m, at the beam center $\psi_e = \psi_a = 0$ thus is

$$m = \frac{d\theta_a}{d\theta_e} = \frac{\cos\theta_e \cos\theta_a}{\sin\theta_e \sin\theta_a} = \frac{1}{\tan\theta_e \tan\theta_a} \qquad (4.24a)$$

and the angle of the tangent line is

$$\phi_{\min} = \arctan(m) \qquad (4.24b)$$

Figure 4.11 is a graph of the normalized bandwidth versus ϕ, the rotation angle of the ellipse shown in Fig. 4.2 for $\sigma_1 = 2$ degrees, $\sigma_2 = 1$ degree, $\theta_a = 60$ degrees, and $\theta_e = 30$ degrees. The minimum in the graph is at 45.1 degree. This value should be approximately that given by Eqs. (4.24)

$$m = \frac{1}{\tan 30 \tan 60} = 1 \qquad (4.25)$$

so that $\phi_{min} = \arctan(1) = 45$ degrees, which is a good approximation to the minimum observed, 45.1 degrees. Note that the maximum then is at $\phi_{max} = \phi_{min} + 90$ degrees.

Chapter 5

Précis of Waveform Analysis Techniques

THE TERRAIN has many randomly distributed scatterers. The Doppler return is the sum of the reflections from these scatterers as they pass through the antenna beam. In consequence, the Doppler radar echo will be modeled as a random waveform whose exact analysis requires the use of generalized harmonic analysis. In order to establish a common notation and viewpoint, the following is a précis of the elements of harmonic and system analysis we'll require.

I.   Harmonic Analysis

The study of systems that interact with complex waveforms is often simplified by expressing the complex waveform as a linear combination of other waveforms from a set with certain specified characteristics. One of the most useful sets for use in the study of linear time-invariant (LTI) systems is the set of sinusoids. The reason is that the response of an LTI system to an input sinusoid is a sinusoid with the same frequency. The representation of a waveform as the sum of sinusoids is called Fourier analysis. If the waveform is periodic, the representation is called a Fourier series; if the waveform is aperiodic, the representation is called a Fourier transform.

A.   The Fourier Series

A periodic waveform $p(t)$ is one that is equal to a shift of itself. That is, there is a number T such that $p(t + T) = p(t)$. Observe that a periodic waveform $p(t)$ must be nonzero for $-\infty < t < \infty$. The number T is called a period of the waveform $p(t)$. Clearly, if T is a period, then so is nT in which $n = 1, 2, 3, \ldots$. The smallest value of $T = T_1$ is called the fundamental period of $p(t)$.

If a waveform $p(t)$ is periodic with a fundamental period T_1, then, with certain mild conditions that are met by physical waveforms, it can be expressed as the linear combination of sinusoids with the harmonic frequencies $\omega_n = n\omega_1$ in which $\omega_1 = 2\pi/T_1$ as

$$p(t) = \sum_{n=0}^{\infty} A_n \cos(\omega_n t) + B_n \sin(\omega_n t) \tag{5.1}$$

This expression is called the Fourier series of the periodic waveform $p(t)$.

By making use of the identity

$$e^{j\omega_n t} = \cos(\omega_n t) + j\sin(\omega_n t) \tag{5.2}$$

each sinusoid can be expressed as the linear combination of two phasors as:

$$\cos(\omega_n t) = \frac{1}{2}e^{j\omega_n t} + \frac{1}{2}e^{-j\omega_n t} \tag{5.3a}$$

and

$$\sin(\omega_n t) = \frac{1}{2j}e^{j\omega_n t} - \frac{1}{2j}e^{-j\omega_n t} \tag{5.3b}$$

so that the periodic waveform also can be expressed as the linear combination of phasors with the frequencies ω_n as

$$p(t) = \sum_{n=-\infty}^{\infty} C_n e^{j\omega_n t} \tag{5.4}$$

in which for positive values of n

$$C_n = \frac{1}{2}(A_n - jB_n) \quad \text{and} \quad C_{-n} = \frac{1}{2}(A_n + jB_n) \tag{5.5}$$

The coefficients C_n are determined by evaluating the integral

$$C_n = \frac{1}{T_1}\int_0^{T_1} p(t)e^{-j\omega_n t}dt \tag{5.6}$$

B. The One-Dimensional Fourier Transform[1]

However, our concern is not with periodic waveforms but with aperiodic ones. Although an aperiodic waveform $f(t)$ cannot be expressed as the sum of phasors as in Eq. (5.4), it can, under certain conditions, be expressed as an integral of phasors as

$$f(t) = \frac{1}{2\pi}\int_{-\infty}^{\infty} F(j\omega)e^{j\omega t}d\omega \tag{5.7}$$

---

[1] See Schetzen, M. *Linear Time-Invariant Systems*, IEEE Press/John Wiley & Sons, NY, NY, 2002 for a more complete discussion of the Fourier transform and its properties.

in which $F(j\omega)$ is the amplitude density of the phasors determined by the integral

$$F(j\omega) = \int_{-\infty}^{\infty} f(t)e^{-j\omega t}\,dt \qquad (5.8)$$

The amplitude density of the phasors $F(j\omega)$ is called the Fourier transform of $f(t)$. A sufficient (but not necessary) condition for the validity of these relations is that $|f(t)| < \infty$ and that

$$I = \int_{-\infty}^{\infty} |f(t)|\,dt < \infty \qquad (5.9)$$

If this condition is satisfied, the two relations, Eq. (5.7) and Eq. (5.8), are called a Fourier transform pair. They are a pair because, if one is true, then so is the other. For example, let $F(j\omega)$ be determined using a given function $f(t)$ in Eq. (5.8). Then if $F(j\omega)$ were used in Eq. (5.7), the integration would result in the same function $f(t)$ used in Eq. (5.8) to determine $F(j\omega)$.[2] The function $f(t)$ is thus called the inverse Fourier transform of $F(j\omega)$. We shall assume condition (5.9) is satisfied for the Fourier transform theory we shall discuss.

As a simple example, let

$$f(t) = e^{-\alpha t}u(t) \qquad (5.10)$$

in which $\alpha > 0$ and $u(t)$ is the unit step function

$$u(t) = \begin{cases} 0 & \text{if } t < 0 \\ 1/2 & \text{if } t = 0 \\ 1 & \text{if } t > 0 \end{cases} \qquad (5.11)$$

Then, from Eq. (5.8)

$$F(j\omega) = \int_{0}^{\infty} e^{-\alpha t}e^{-j\omega t}\,dt$$

$$= \int_{0}^{\infty} e^{-(\alpha + j\omega)t}\,dt = \frac{1}{\alpha + j\omega} \qquad (5.12)$$

Because $f(t)$ of our example satisfies the condition given by Eq. (5.9), we have that Eq. (5.7) and Eq. (5.8) are a Fourier transform pair. We thus are guaranteed that

$$\frac{1}{2\pi}\int_{-\infty}^{\infty} \frac{1}{\alpha + j\omega}e^{j\omega t}\,d\omega = \begin{cases} 0 & \text{for } t < 0 \\ 1/2 & \text{for } t = 0 \\ e^{-at} & \text{for } t > 0 \end{cases} = f(t) \qquad (5.13)$$

---

[2] More precisely, at points of discontinuity of $f(t)$, it is equal to the midpoint value at the discontinuity. This is the reason I define the value of $f(t)$ at a discontinuity to be equal to the value at the discontinuity midpoint as in Eq. (5.11).

The value of the integral can be verified by direct integration or, more simply, from a table of definite integrals.

Many properties of Fourier transforms have been derived, which greatly simplifies the determination of the Fourier transform and its inverse. The properties also lend a great deal of insight and understanding of the results obtained. For example, the Fourier transform of

$$g(t) = f(t)e^{-j\beta t} \tag{5.14a}$$

is

$$
\begin{aligned}
G(j\omega) &= \int_0^\infty g(t)e^{-j\omega t}\,dt \\
&= \int_0^\infty f(t)e^{-j\beta t}e^{-j\omega t}\,dt \\
&= \int_0^\infty f(t)e^{-j(\omega+\beta)t}\,dt \\
&= F[j(\omega+\beta)]
\end{aligned}
\tag{5.14b}
$$

As an example of this result, let

$$g(t) = f(t)\cos(\omega_0 t) \tag{5.15}$$

Using the identity

$$\cos(\omega t) = \frac{1}{2}e^{j\omega t} + \frac{1}{2}e^{-j\omega t} \tag{5.16}$$

we can express $g(t)$ in the form

$$g(t) = \frac{1}{2}f(t)e^{j\omega_0 t} + \frac{1}{2}f(t)e^{-j\omega_0 t} \tag{5.17}$$

By the use of Eqs 5.14 we then immediately have

$$G(j\omega) = \frac{1}{2}F[j(\omega-\omega_0)] + \frac{1}{2}F[j(\omega+\omega_0)] \tag{5.18}$$

As an illustration, with $f(t)$ given by Eq. (5.10), we then have

$$g(t) = e^{-\alpha t}\cos(\omega_0 t)u(t) \tag{5.19}$$

so that, with the result given by Eq. (5.12), we obtain

$$
\begin{aligned}
G(j\omega) &= \frac{1}{2}\frac{1}{\alpha + j(\omega - \omega_0)} + \frac{1}{2}\frac{1}{\alpha + j(\omega + \omega_0)} \\
&= \frac{\alpha + j\omega}{[\alpha + j(\omega + \omega_0)][\alpha + j(\omega - \omega_0)]}
\end{aligned}
\tag{5.20}
$$

In addition to simplifying calculations as above, a great deal of insight into Fourier transforms can be obtained from their various properties. The properties we need will be presented later in this précis in the context in which they'll be used so that they can be interpreted in a manner that is physically meaningful in the application.

However, there is one other property that we'll obtain here. The property is that the Fourier transform of the conjugate of $f(-t)$ is the conjugate of $F(j\omega)$. To obtain this property, we have from Eq. (5.8) that $F^*(j\omega)$, the conjugate of $F(j\omega)$, is

$$
F^*(j\omega) = \left[\int_{-\infty}^{\infty} f(t)e^{-j\omega t}\,dt\right]^* = \int_{-\infty}^{\infty} f^*(t)e^{+j\omega t}\,dt
\tag{5.21}
$$

Thus, with the change of variable $\tau = -t$ in the integral we obtain

$$
F^*(j\omega) = \int_{-\infty}^{\infty} f^*(-\tau)e^{-j\omega\tau}\,d\tau
\tag{5.22}
$$

We thus observe that the Fourier transform of $f^*(-t)$ is $F^*(j\omega)$. As an example of the use of this property, we'll determine the Fourier transform of

$$
s(t) = e^{-\alpha|t|} \quad \text{in which } \alpha > 0
\tag{5.23}
$$

With the use of the step function defined by Eq. (5.11), the function $s(t)$ can be written in the form

$$
\begin{aligned}
s(t) &= e^{-\alpha t}u(t) + e^{\alpha t}u(-t) \\
&= f(t) + f(-t)
\end{aligned}
\tag{5.24}
$$

in which $f(t)$ is the exponential time function given by Eq. (5.10). We then have that the Fourier transform of $s(t)$ is

$$
\begin{aligned}
S(j\omega) &= F(j\omega) + F^*(j\omega) \\
&= \frac{1}{\alpha + j\omega} + \frac{1}{\alpha - j\omega} \\
&= \frac{2\alpha}{\alpha^2 + \omega^2}
\end{aligned}
\tag{5.25}
$$

C.   The Two-Dimensional Fourier Transform[3]

In our spectral analysis of the radar Doppler echo, we need the extension of the one-dimensional Fourier transform pair relations, Eq. (5.7) and Eq. (5.8), to two dimensions. We shall develop this before proceeding. For this, consider a function $f_2(t_1, t_2)$ for which $|f_2(t_1, t_2)| < \infty$ and

$$I = \int_{-\infty}^{\infty} \int_{-\infty}^{\infty} |f_2(t_1, t_2)| dt_1 dt_2 < \infty \tag{5.26}$$

This condition is equivalent to condition (5.9) for the one-dimensional case. It is a condition that is sufficient but not necessary to assure the convergence of all integrals.

The two-dimensional Fourier transform is obtained by determining the transform of $f_2(t_1, t_2)$ with respect to one variable at a time. By holding the variable t_2 constant, the one-dimensional Fourier transform is

$$F_1(j\omega_1, t_2) = \int_{-\infty}^{\infty} f_2(t_1, t_2) e^{-j\omega_1 t_1} dt_1 \tag{5.27}$$

Now by holding the variable ω_1 constant, the one-dimensional Fourier transform of $F_1(j\omega_1, t_2)$ with respect to t_2 is

$$\begin{aligned}
F_2(j\omega_1, j\omega_2) &= \int_{-\infty}^{\infty} F_1(j\omega_1, t_2) e^{-j\omega_2 t_2} dt_2 \\
&= \int_{-\infty}^{\infty} \int_{-\infty}^{\infty} f_2(t, t_2) e^{-j\omega_1 t_1} e^{-j\omega_2 t_2} dt_1 dt_2
\end{aligned} \tag{5.28}$$

This last equation is obtained by substituting Eq. (5.27). The function $F_2(j\omega_1, j\omega_2)$ is called the two-dimensional Fourier transform of $f_2(t_1, t_2)$.

The inverse transform can now be obtained by reversing our process above. First, by use of Eq. (5.7), we have from Eq. (5.28)

$$F_1(j\omega_1, t_2) = \frac{1}{2\pi} \int_{-\infty}^{\infty} F_2(j\omega_1, j\omega_2) e^{j\omega_2 t_2} d\omega_2 \tag{5.29}$$

---

[3] For a more complete discussion of the two-dimensional Fourier transform and its properties, see Schetzen, M. *The Volterra & Wiener Theories of Nonlinear Systems*, J. Wiley & Sons, New York, N.Y., 1980, Reprinted with additional material by R.E. Krieger Publishing Co., Malabar, FL, 1989. An updated reprint with new additional material will be available from R. E. Krieger Publishing Co., Malabar, Fl., in 2006.

and from Eq. (5.27)

$$f_2(t_1, t_2) = \frac{1}{2\pi} \int_{-\infty}^{\infty} F_1(j\omega_1, t_2) e^{j\omega_1 t_1} d\omega_1$$

$$= \frac{1}{(2\pi)^2} \int_{-\infty}^{\infty} \int_{-\infty}^{\infty} F_2(j\omega_1, j\omega_2) e^{j\omega_1 t_1} e^{j\omega_2 t_2} d\omega_1 d\omega_2 \qquad (5.30)$$

The last equation is obtained by substituting Eq. (5.29). Equation (5.30) is the inverse transform of $F_2(j\omega_1, \omega_2)$. We thus have established the two-dimensional Fourier transform pair

$$F_2(j\omega_1, j\omega_2) = \int_{-\infty}^{\infty} \int_{-\infty}^{\infty} f_2(t, t_2) e^{-j(\omega_1 t_1 + \omega_2 t_2)} dt_1 dt_2 \qquad (5.31a)$$

and

$$f_2(t_1, t_2) = \frac{1}{(2\pi)^2} \int_{-\infty}^{\infty} \int_{-\infty}^{\infty} F_2(j\omega_1, j\omega_2) e^{j(\omega_1 t_1 + \omega_2 t_2)} d\omega_1 d\omega_2 \qquad (5.31b)$$

As a simple example, we'll determine the two-dimensional Fourier transform of

$$f(t_1, t_2) = e^{-(\alpha_1 t_1 + \alpha_2 t_2)} u(t_1) u(t_2) \qquad (5.32)$$

in which $\alpha_1 > 0$ and $\alpha_2 > 0$. Substituting this function in Eq. (5.31a), we obtain

$$F_2(j\omega_1, j\omega_2) = \int_0^{\infty} \int_0^{\infty} e^{-(\alpha_1 t_1 + \alpha_2 t_2)} e^{-j(\omega_1 t_1 + \omega_2 t_2)} dt_1 dt_2$$

$$= \int_0^{\infty} \int_0^{\infty} e^{-(\alpha_1 + j\omega_1)t_1} e^{-(\alpha_2 + j\omega_2)t_2} dt_1 dt_2$$

$$= \int_0^{\infty} e^{-(\alpha_1 + j\omega_1)t_1} dt_1 \int_0^{\infty} e^{-(\alpha_2 + j\omega_2)t_2} dt_2$$

$$= \frac{1}{(\alpha_1 + j\omega_1)(\alpha_2 + j\omega_2)} \qquad (5.33)$$

Because $f(t_1, t_2)$ of our example satisfies the condition given by Eq. (5.26), we have that Eq. (5.31a) and Eq. (5.31b) are a Fourier transform pair. We thus are guaranteed that

$$\frac{1}{(2\pi)^2} \int_{-\infty}^{\infty} \int_{-\infty}^{\infty} \frac{1}{(\alpha_1 + j\omega_1)(\alpha_2 + j\omega_2)} e^{j(\omega_1 t_1 + \omega_2 t_2)} d\omega_1 d\omega_2$$

$$= e^{-(\alpha_1 t_1 + \alpha_2 t_2)} u(t_1) u(t_2) \qquad (5.34)$$

The calculation of the two-dimensional Fourier transform for this example was not difficult because $f(t_1, t_2)$ can be expressed as

$$f(t_1, t_2) = f_1(t_1)f_2(t_2) \tag{5.35a}$$

in which

$$f_1(t_1) = e^{-\alpha_1 t_1} u(t_1) \tag{5.35b}$$

and

$$f_2(t_2) = e^{-\alpha_2 t_2} u(t_2) \tag{5.35c}$$

so that the two-dimensional Fourier transform for the example could be viewed as the product of two one-dimensional Fourier transforms. Generally, when this is not the case, the evaluation of the two-dimensional Fourier transform is not so simple.

Our analysis of the Doppler spectrum will involve the two-dimensional Fourier transform. However, we'll require only certain properties of the Doppler spectrum such as the spectrum center frequency and the spectrum bandwidth. The determination of those properties will be obtained from the two-dimensional Fourier transform expressions without having to evaluate them. However, the numerical values of spectrum center frequency and bandwidth will require the evaluation of some two-dimensional integrals. A computer program was written for their evaluation on a standard computer. Graphs of the spectrum bandwidth and center frequency are also generated by the computer program.

II.   LTI System Concepts

The theory of LTI systems is a mature theory that has been well developed.[4] For our précis, only those results we shall use will be discussed. To begin, a system depicted in Fig. 5.1 is simply something for which there is a rule by which input waveforms $x(t)$ are mapped into output waveforms $y(t)$.

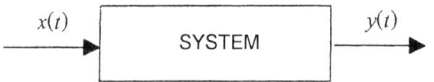

Fig. 5.1   Representation of a system.

---

[4] For a complete discussion of LTI systems, see Schetzen, M. *Linear Time-Invariant Systems,* IEEE Press/John Wiley & Sons, NY, NY, 2002.

A. The Time Domain

A system is time invariant (TI) if the rule does not vary with time. A system is linear if, for any constants C_1 and C_2 and for any inputs $x_1(t)$ and $x_2(t)$, the input $x(t) = C_1 x_1(t) + C_2 x_2(t)$ results in the output $y(t) = C_1 y_1(t) + C_2 y_2(t)$ in which $y_1(t)$ is the output resulting from the input $x_1(t)$ and $y_2(t)$ is the output resulting from the input $x_2(t)$. This is called generalized superposition so that a linear system is one that satisfies generalized superposition. A system that is both linear and TI is referred to as an LTI system.

We shall only be interested in LTI systems that are bounded-input bounded-output (BIBO) stable. A system is BIBO stable if every bounded input results in an output that also is bounded. A waveform $x(t)$ is bounded if $|x(t)| \leq M$ for some value M. That is, a waveform is bounded if its magnitude never exceeds some value M at any time. A BIBO-stable system thus is one that is not explosive because no bounded input waveform of the system will produce an output waveform that is infinite at one or more instants of time.

If an LTI system is BIBO stable, then it can be shown that the system output can be determined from the system input by the convolution integral,

$$y(t) = h(t) * x(t) = \int_{-\infty}^{\infty} h(\sigma)x(t - \sigma)d\sigma \qquad (5.36)$$

In this expression, the function $h(t)$ is called the system unit-impulse response because it is the LTI system response to the input $x(t) = \delta(t)$. The function $\delta(t)$ is called the unit impulse. As shown in Fig. 5.2a, the unit impulse is a positive pulse that is centered at $t = 0$; its width δ is infinitesimal (not zero) and its area is one; its usual representation is shown in Fig. 5.2b. Note then that $h(t)$ is a real function of t if the system is a physical one.

Because the output can be determined for any input with the use of Eq. (5.36), we observe that the input-output mapping of a BIBO-stable LTI system is completely determined once the system unit-impulse response is known.

It can be shown that an LTI system is BIBO stable if and only if

$$I = \int_{-\infty}^{\infty} |h(t)|dt < \infty \qquad (5.37)$$

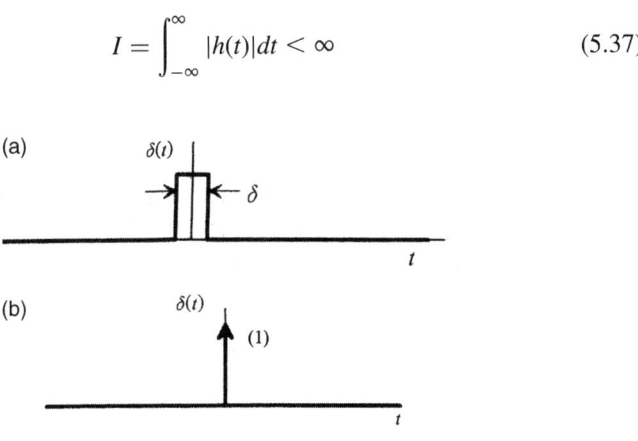

Fig. 5.2   Representation of the unit impulse.

B. The Frequency Domain

The phasor $e^{j\omega t}$ is an important waveform in LTI system theory because if the input is a phasor, then the output is a constant times the input phasor. In mathematical terms, the phasor is called an eigenfunction of the system mapping. To see this, we let $x(t) = e^{j\omega t}$ in the convolution integral Eq. (5.36) to obtain

$$y(t) = \int_{-\infty}^{\infty} h(\sigma)e^{j\omega(t-\sigma)}d\sigma = e^{j\omega t}\int_{-\infty}^{\infty} h(\sigma)e^{-j\omega\sigma}\,d\sigma \qquad (5.38)$$

so that

$$y(t) = H(j\omega)e^{j\omega t} \qquad (5.39)$$

in which

$$H(j\omega) = \int_{-\infty}^{\infty} h(\sigma)e^{-j\omega\sigma}\,d\sigma \qquad (5.40)$$

The function $H(j\omega)$ is called the transfer function of the LTI system. The input is bounded because $|x(t)| = |e^{j\omega t}| = 1$ and the system is BIBO stable, thus the output must be bounded so that

$$|y(t)| = |H(j\omega)e^{j\omega t}| = |H(j\omega)||e^{j\omega t}| = |H(j\omega)| < \infty \qquad (5.41)$$

We thus observe that the transfer function is bounded.

A physical interpretation of the transfer function can be obtained in terms of the system response to an input that is a sinusoid. For this, let the input be

$$x(t) = A\cos(\omega t + \phi) \qquad (5.42)$$

By expressing $x(t)$ as the linear combination of phasors as

$$x(t) = \frac{1}{2}Ae^{j\phi}e^{j\omega t} + \frac{1}{2}Ae^{-j\phi}e^{-j\omega t} \qquad (5.43)$$

and using the linearity property, generalized superposition, and Eq. (5.39), the corresponding output can be shown to be

$$y(t) = A|H(j\omega)|\cos[\omega t + \phi + \theta(\omega)] \qquad (5.44a)$$

in which

$$\theta(\omega) = \sphericalangle H(j\omega) \qquad (5.44b)$$

The output is observed to be a sinusoid with the same frequency as the input sinusoid. However, the output amplitude equals the input amplitude times $|H(j\omega)|$. In consequence, the magnitude of the transfer function $|H(j\omega)|$ is called the system gain. Also, the phase of the output sinusoid equals the phase of the input sinusoid plus $\sphericalangle H(j\omega)$. In consequence, the angle of the transfer function is called the system phase shift. Note that the system transfer function can be determined experimentally by measuring the system gain and phase shift.

Observe from Eq. (5.40) that the transfer function is simply the Fourier transform of the LTI system unit-impulse response. Because the system is BIBO stable, we have that Eq. (5.37) is satisfied. Thus, from Fourier transform theory, $h(t)$ can be obtained as the inverse Fourier transform of the transfer function as

$$h(t) = \frac{1}{2\pi} \int_{-\infty}^{\infty} H(j\omega)e^{j\omega t}\,d\omega \qquad (5.45)$$

Consequently, the system unit-impulse response can be determined from the system gain and phase shift. We thus observe that knowledge of the transfer function allows the determination of the output for any input. In consequence, the input-output mapping of a stable LTI system is completely determined once the system transfer function is known.

The transfer function allows the analysis of BIBO-stable LTI systems in the frequency domain. For this we consider inputs $x(t)$ for which

$$\int_{-\infty}^{\infty} |x(t)|dt < \infty \qquad (5.46)$$

so that $X(j\omega)$, the Fourier transform of $x(t)$, exists. The Fourier transform of the output $y(t)$ of the stable LTI system then exists and is

$$Y(j\omega) = \int_{-\infty}^{\infty} y(t)e^{-j\omega t}\,dt \qquad (5.47)$$

By substituting the convolution relation, Eq. (5.36), the Fourier transform of the output $y(t)$ can be expressed in terms of that of the input as

$$Y(j\omega) = H(j\omega)X(j\omega) \qquad (5.48)$$

One important advantage of the frequency domain is that expressions such as Eq. (5.48) are algebraic while the equivalent expressions in the time domain such as Eq. (5.36) are integrals.

C.   A Parseval Relation

An important relation that we shall need is the Parseval relation:

$$\int_{-\infty}^{\infty} f_1(t)f_2^*(t)dt = \frac{1}{2\pi} \int_{-\infty}^{\infty} F_1(j\omega)F_2^*(j\omega)\,d\omega \qquad (5.49)$$

This relation can be obtained by noting from the convolution relation, Eq. (5.36), that $y(t)$ at $t = 0$ is

$$y(0) = \int_{-\infty}^{\infty} h(\sigma)x(-\sigma)d\sigma \tag{5.50a}$$

However, from the Fourier transform Eq. (5.7) and using Eq. (5.48), we have

$$y(0) = \frac{1}{2\pi} \int_{-\infty}^{\infty} Y(j\omega)d\omega = \frac{1}{2\pi} \int_{-\infty}^{\infty} H(j\omega)X(j\omega)d\omega \tag{5.50b}$$

We then obtain by equating Eq. (5.50a) and Eq. (5.50b)

$$\int_{-\infty}^{\infty} h(\sigma)x(-\sigma)d\sigma = \frac{1}{2\pi} \int_{-\infty}^{\infty} H(j\omega)X(j\omega)d\omega \tag{5.51}$$

Now, let

$$h(t) = f_1(t) \quad \text{and} \quad x(t) = f_2^*(-t) \tag{5.52}$$

Then from the Fourier transform property given by Eq. (5.22), we have

$$H(j\omega) = F_1(j\omega) \quad \text{and} \quad X(j\omega) = F_2^*(j\omega) \tag{5.53}$$

By substituting these relations into Eq. (5.51) we obtain the desired Parseval relation, which is

$$\int_{-\infty}^{\infty} f_1(t)f_2^*(t)dt = \frac{1}{2\pi} \int_{-\infty}^{\infty} F_1(j\omega)F_2^*(j\omega)d\omega \tag{5.54a}$$

An important special case of the Parseval relation is that for which $f_1(t) = f(t)$ and also $f_2(t) = f(t)$

$$\int_{-\infty}^{\infty} |f(t)|^2 dt = \frac{1}{2\pi} \int_{-\infty}^{\infty} |F(\omega)|^2 d\omega \tag{5.54b}$$

D. The Energy Density Spectrum

This last relation, Eq. (5.54b), is often called the energy theorem. To understand the reason for this name, let $f(t)$ be the current through a 1-ohm resistor. Then the total energy dissipated in the resistor is given by the left side of Eq. (5.54b). Because the right side of this equation is the integral overall frequencies of $1/2\pi |F(j\omega)|^2$, we can interpret it as an energy density spectrum in joules/radian per second. Equivalently, because one radian per second equals 2π hertz, we can interpret $|F(j\omega)|^2$ as an energy density spectrum in joules/hertz. For this reason, $|F(j\omega)|^2$ is often called the *energy density spectrum* of $f(t)$.

To make the interpretation of an energy density spectrum more concrete, consider an ideal method to measure the energy density spectrum of a waveform $x(t)$. Ideally, to measure the energy of $x(t)$ in the band of frequencies $0 < \omega < \omega_1$, we would apply $x(t)$ to an ideal low-pass filter with the cut-off frequency ω_1. An ideal low-pass filter is one that has unity gain in the passband and zero gain for all frequencies above the cut-off frequency ω_1. That is,

$$|H(j\omega)| = \begin{cases} 1 & \text{for} \quad |\omega| < \omega_1 \\ 0 & \text{for} \quad |\omega| > \omega_1 \end{cases} \tag{5.55}$$

As shown in Fig. 5.3, the output of the ideal low-pass filter is a voltage that is applied to a 1-ohm resistor. In accordance with Eq. (5.48), the Fourier transform of the output waveform $y(t)$ is $Y(j\omega) = H(j\omega)X(j\omega)$. Thus, with the use of the Parseval relation, Eq. (5.54b), the total energy dissipated in the 1-ohm resistor is

$$\int_{-\infty}^{\infty} |y(t)|^2 dt = \frac{1}{2\pi} \int_{-\infty}^{\infty} |Y(j\omega)|^2 d\omega$$

$$= \frac{1}{2\pi} \int_{-\infty}^{\infty} |H(j\omega)X(j\omega)|^2 d\omega$$

$$= \frac{1}{2\pi} \int_{-\infty}^{\infty} |H(j\omega)|^2 |X(j\omega)|^2 d\omega \tag{5.56}$$

Thus, with the use of Eq. (5.55), the total energy dissipated in the resistor is

$$\int_{-\infty}^{\infty} |y(t)|^2 dt = \frac{1}{2\pi} \int_{-\omega_1}^{\omega_1} |X(j\omega)|^2 d\omega \tag{5.57}$$

Similarly, to measure the total energy contained by the waveform $x(t)$ in the frequency band $\omega_1 < \omega < \omega_2$, we would use an ideal bandpass filter with unity gain in the given band and zero gain outside the given band. In accordance

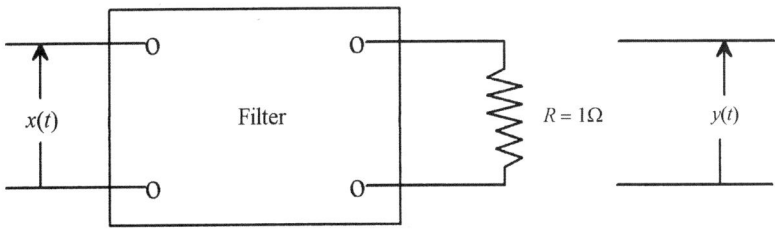

Fig. 5.3   The Measurement of an Energy Density Spectrum.

with Eq. (5.56), the total energy dissipated by the 1-ohm resistor is then

$$\int_{-\infty}^{\infty} |y(t)|^2 \, dt = \frac{1}{2\pi} \int_{-\omega_2}^{-\omega_1} |X(j\omega)|^2 \, d\omega + \frac{1}{2\pi} \int_{\omega_1}^{\omega_2} |X(j\omega)|^2 \, d\omega \qquad (5.58)$$

If $x(t)$ is a real function of t, then $|X(j\omega)|$ is an even function of ω. Thus, for the physical case in which $x(t)$ is a real function of t, both integrals on the right side of Eq. (5.58) have the same value so that we also can express Eq. (5.58) in the form

$$\int_{-\infty}^{\infty} |y(t)|^2 \, dt = \frac{2}{2\pi} \int_{\omega_1}^{\omega_2} |X(j\omega)|^2 \, d\omega \qquad (5.59)$$

Using the relation $\omega = 2\pi f$, this expression also can be written as

$$\int_{-\infty}^{\infty} |y(t)|^2 \, dt = 2 \int_{f_1}^{f_2} |X(j2\pi f)|^2 \, df \qquad (5.60)$$

Observe that the experimental measurement of the energy contained by a real time function $x(t)$ in a given frequency band $f_1 < f < f_2$ hertz is numerically equal to twice the area under $|X(j2\pi f)|^2$ in the given band. That is, $|X(j2\pi f)|^2$ is the actual energy density spectrum of $x(t)$ in joules per hertz that we would ideally measure in the laboratory.

E. The Parseval Relation in Two Dimensions

In our development of a theory for the spectral analysis of the radar Doppler echo, we shall have need for an extension of the Parseval relation to two dimensions. For this, we consider two functions, $f_a(t_1, t_2)$, and $f_b(t_1, t_2)$. To ensure the convergence of all integrals, we require that each function satisfy Eq. (5.26). Now consider the integral

$$I = \int_{-\infty}^{\infty} \int_{-\infty}^{\infty} f_a(t_1, t_2) f_b^*(t_1, t_2) \, dt_1 \, dt_2 \qquad (5.61)$$

As previously, the star in the integral indicates the conjugate of the function. To obtain the desired Parseval relation, we take the conjugate of the two-dimensional Fourier transform pair relation, Eq. (5.31b),

$$f_b^*(t_1, t_2) = \frac{1}{(2\pi)^2} \int_{-\infty}^{\infty} \int_{-\infty}^{\infty} F_b^*(j\omega_1, j\omega_2) \, e^{-j(\omega_1 t_1 + \omega_2 t_2)} \, d\omega_1 \, d\omega_2 \qquad (5.62)$$

Now substitute this expression in Eq. (5.61)

$$I = \frac{1}{(2\pi)^2} \int_{-\infty}^{\infty} \int_{-\infty}^{\infty} f_a(t_1, t_2) dt_1 dt_2$$

$$\int_{-\infty}^{\infty} \int_{-\infty}^{\infty} F_b^*(j\omega_1, j\omega_2) e^{-j(\omega_1 t_1 + \omega_2 t_2)} d\omega_1 d\omega_2 \qquad (5.63)$$

In this equation, the value of I is obtained by first integrating with respect to ω_1 and ω_2. The result of this integration is $f_b^*(t_1, t_2)$, which is then inserted in the first set of double integrals and integrated. Because the condition given by Eq. (5.26) is satisfied, we can interchange the order of integration to obtain

$$I = \frac{1}{(2\pi)^2} \int_{-\infty}^{\infty} \int_{-\infty}^{\infty} F_b^*(j\omega_1, j\omega_2) d\omega_1 d\omega_2$$

$$\int_{-\infty}^{\infty} \int_{-\infty}^{\infty} f_a(t_1, t_2) e^{-j(\omega_1 t_1 + \omega_2 t_2)} dt_1 dt_2 \qquad (5.64)$$

In this equation, the value of I is obtained by first integrating with respect to t_1 and t_2. From Eq. (5.31a), the result of this integration is

$$F_a(j\omega_1, j\omega_2) = \int_{-\infty}^{\infty} \int_{-\infty}^{\infty} f_a(t_1, t_2) e^{-j(\omega_1 t_1 + \omega_2 t_2)} dt_1 dt_2 \qquad (5.65)$$

The function $F_a(j\omega_1, j\omega_2)$ is then inserted in the first set of double integrals and integrated to obtain

$$I = \frac{1}{(2\pi)^2} \int_{-\infty}^{\infty} \int_{-\infty}^{\infty} F_a(\omega_1, \omega_2) F_b^*(\omega_1, \omega_2) d\omega_1 d\omega_2 \qquad (5.66)$$

The expressions given by Eq. (5.61) and (5.66) are equal because they are both equal to I. We thus have

$$\int_{-\infty}^{\infty} \int_{-\infty}^{\infty} f_a(t_1, t_2) f_b^*(t_1, t_2) dt_1 dt_2$$

$$= \frac{1}{(2\pi)^2} \int_{-\infty}^{\infty} \int_{-\infty}^{\infty} F_a(\omega_1, \omega_2) F_b^*(\omega_1, \omega_2) d\omega_1 d\omega_2 \qquad (5.67)$$

This is the Parseval relation in two dimensions that we'll need in our development of a theory for the spectral analysis of the radar Doppler echo. Note that for the special case in which $f_b(t_1, t_2) = f_a(t_1, t_2)$, we have

$$\int_{-\infty}^{\infty} \int_{-\infty}^{\infty} |f_a(t_1, t_2)|^2 dt_1 dt_2 = \frac{1}{(2\pi)^2} \int_{-\infty}^{\infty} \int_{-\infty}^{\infty} |f_a(\omega_1, \omega_2)|^2 d\omega_1 d\omega_2 \qquad (5.68)$$

This is the two-dimensional extension of Eq. (5.54b).

III. Continuing Time Functions

The Fourier transforms of time functions, $f(t)$, we have discussed to this point are valid only for those for which the integral of its magnitude is finite so that they satisfy Eq. (5.9). However, the Doppler radar echo is not such a waveform. For its analysis, we consider it to be a waveform that has existed for a very long time, in fact an infinite time so that its existence is for $-\infty < t < \infty$. Such waveforms are called continuing waveforms because they continue forever. Our objective is to determine the frequency spectrum of such a waveform. Now, the mean-square value of a waveform $f_a(t)$ is[5]

$$\overline{f_a^2(t)} = \lim_{T \to \infty} \frac{1}{2T} \int_{-T}^{T} f_a^2(t) dt \qquad (5.69)$$

Clearly, the mean-square value of normal continuing waveforms is greater than zero. It can be shown that the Fourier transform of any waveform with a nonzero mean-square value does not exist.[6] This places a difficulty in our determination of the waveform frequency spectrum because Fourier transform theory cannot be used. We can, however, determine the waveform power density spectrum by determining its autocorrelation function.

A. The Autocorrelation Function

The autocorrelation function of a real continuing time function $f_a(t)$ is

$$\phi_{aa}(\tau) = \overline{f_a(t)f_a(t + \tau)} = \lim_{T \to \infty} \frac{1}{2T} \int_{-T}^{T} f_a(t)f_a(t + \tau) \, dt \qquad (5.70)$$

For $\tau = 0$, we observe that $\phi_{aa}(0) = \overline{f_a^2(t)}$. There are some important properties of the autocorrelation function. One property is that it is an even function of τ. To show this, we have from our definition of the autocorrelation function that

$$\phi_{aa}(-\tau) = \lim_{T \to \infty} \frac{1}{2T} \int_{-T}^{T} f_a(t)f_a(t - \tau) \, dt \qquad (5.71a)$$

---

[5] The overbar will be used to indicate the time average as in Eq. (5.69).
[6] A periodic waveform is a special case of a continuing waveform. For its harmonic analysis, Fourier series is used. A subterfuge for representing periodic waveforms using Fourier transforms is to use impulses located at the harmonic frequencies. As discussed in Footnote 7, this subterfuge cannot be used for continuing aperiodic waveforms such as the Doppler radar echo.

We now make the change of variable $\sigma = t - \tau$ to obtain

$$\phi_{aa}(-\tau) = \lim_{T \to \infty} \frac{1}{2T} \int_{-T-\tau}^{T-\tau} f_a(\sigma + \tau) f_a(\sigma) \, dt \qquad (5.71b)$$

The interval of integration is $(-T - \tau, T - \tau)$. The interval becomes $(-\infty, \infty)$ in the limit as $T \to \infty$. We thus can write

$$\phi_{aa}(-\tau) = \lim_{T \to \infty} \frac{1}{2T} \int_{-T}^{T} f_a(\sigma + \tau) f_a(\sigma) \, dt \qquad (5.71c)$$

By comparing this expression with Eq. (5.70) we have

$$\phi_{aa}(-\tau) = \phi_{aa}(\tau) \qquad (5.72)$$

so that the autocorrelation function is an even function.

A second important property is that the absolute maximum value of the auto-correlation function is at $\tau = 0$. To show this, consider the function

$$g_a(t) = f_a(t) \pm f_a(t + \tau) \qquad (5.73a)$$

The mean-square value of $g_a(t)$ is

$$\begin{aligned} \overline{g_a^2(t)} &= \overline{[f_a(t) \pm f_a(t + \tau)]^2} = \overline{f_a^2(t)} + \overline{f_a^2(t + \tau)} \pm 2\overline{f_a(t) f_a(t + \tau)} \\ &= \phi_{aa}(0) + \phi_{aa}(0) \pm 2\phi_{aa}(\tau) \\ &= 2[\phi_{aa}(0) \pm \phi_{aa}(\tau)] \end{aligned} \qquad (5.73b)$$

Because $\overline{g_a^2(t)} \geq 0$, we observe that we require for $\tau \neq 0$

$$\phi_{aa}(0) \geq \pm\phi_{aa}(\tau) \qquad (5.73c)$$

This result also can be expressed as

$$\phi_{aa}(0) \geq |\phi_{aa}(\tau)| \qquad (5.73d)$$

Observe from Eq. (5.73b) that $\overline{g_a^2(t)} = 0$ only if $f_a(t)$ is periodic and τ is a period of $f_a(t)$. If we consider only aperiodic functions, then we eliminate the possibility of the equality and so, for continuing aperiodic functions such as the Doppler radar echo, we have the strict inequality

$$\phi_{aa}(0) > |\phi_{aa}(\tau)|, \quad \tau \neq 0 \qquad (5.74)$$

Thus the maximum value of the autocorrelation function of a real aperiodic con-tinuing function is $\phi_{aa}(0) = \overline{f_a^2(t)}$ and for no other value of τ can the magnitude of $\phi_{aa}(\tau)$ equal $\phi_{aa}(0)$.

B. The Power Density Spectrum

The importance of the autocorrelation function for us arises from the Wiener theorem for autocorrelation. Wiener had developed a theory of quadratic variation, which is a Parseval-like relation relating the mean-square value of a waveform to the integral of the square of a type of transform $G(\omega)$, called the integrated Fourier transform.[7] The theorem of quadratic variation he proved is

$$\lim_{T \to \infty} \frac{1}{2T} \int_{-T}^{T} f_a^2(t)dt = \lim_{\varepsilon \to 0} \frac{1}{2\varepsilon} \int_{-\infty}^{\infty} |G(\omega + \varepsilon) - G(\omega - \varepsilon)|^2 \, d\omega \qquad (5.75)$$

Then, by using the algebraic identity

$$4f_a(t)f_a(t + \tau) = [f_a(t) + f_a(t + \tau)]^2 - [f_a(t) - f_a(t + \tau)]^2 \qquad (5.76)$$

he expressed the autocorrelation function as the difference of two mean-square values.[8] With the use of his theory of quadratic variation he was then able to obtain his theorem for autocorrelation

$$\phi_{aa}(\tau) = \lim_{\varepsilon \to 0} \frac{1}{2\varepsilon} \int_{-\infty}^{\infty} |G_a(\omega + \varepsilon) - G_a(\omega - \varepsilon)|^2 \cos(\omega\tau) \, d\omega \qquad (5.77)$$

From this result, Wiener showed that the Fourier transform of the autocorrelation function does exist and is an even and positive function of frequency, which is the power density spectrum of the time function $f_a(t)$.[9] We thus can write

$$\Phi_{aa}(\omega) = \frac{1}{2\pi} \int_{-\infty}^{\infty} \phi_{aa}(\tau)e^{-j\omega\tau} \, d\tau \qquad (5.78a)$$

---

[7] The integrated Fourier transform $G(\omega)$ obtains it name from the result that if the Fourier transform $F(j\omega)$ of a function exists, then the derivative of $G(\omega)$ is equal to $F(j\omega)$. For periodic functions, $G(\omega)$ is a step function with the steps occurring at the harmonic frequencies. It is from this that we are able to define the Fourier transform of a periodic function by using impulses located at its harmonic frequencies with areas equal to the amount of the steps. Generally, $G(\omega)$ is not differentiable at frequencies for which $f(t)$ contains nonzero average power density. At such frequencies, $G(\omega)$ is continuous but not differentiable. An important class of waveforms are those for which there is no frequency interval in which the average power density is zero. The future of such waveforms cannot be predicted with arbitrarily small error from its past; information-bearing waveforms are such waveforms. For these important waveforms $G(\omega)$ is one of those peculiar functions that are continuous everywhere but differentiable nowhere.

[8] This identity has often been used to construct analog multipliers by using two square-law devices.

[9] A good résumé of the proof of the Wiener theorem of autocorrelation is contained in Wiener, N. *Extrapolation, Interpolation, and Smoothing of Stationary Time Series*, The Technoogy Press of M.I.T. & J. Wiley & Sons Inc, New York, 1957, pp. 37–43 and in Lee, Y. W., *Statistical Theory of Communication*, John Wiley & Sons Inc., New York, 1960, pp. 93–96.

and its inverse Fourier transform

$$\phi_{aa}(\tau) = \int_{-\infty}^{\infty} \Phi_{aa}(j\omega)e^{j\omega\tau}d\omega \tag{5.78b}$$

It is possible for the autocorrelation function to be the sum of an aperiodic and a periodic function. If so, we represent the Fourier transform of the periodic function by using impulses located at the harmonic frequencies. Thus, although the Fourier transform of an aperiodic continuing function does not exist, the Fourier transform of its autocorrelation function does exist. From the Wiener theorem of autocorrelation, Eqs. (5.78), we observe that the Fourier transform of the autocorrelation function is a real, even, and positive function of frequency. The autocorrelation function thus is the inverse Fourier transform of a real, even, and positive function. With this result, many specific properties of the autocorrelation function can be obtained.[10]

We can show that the Fourier transform of the autocorrelation function $\Phi_{aa}(\omega)$ is indeed the power density spectrum of the continuing waveform that one would experimentally measure in the laboratory. For this, consider the LTI system depicted in Figure 5.1 with the input $x(t)$ and response $y(t)$. With the use of Eq. (5.36), the autocorrelation of the system output is

$$\phi_{yy}(\tau) = \overline{y(t)y(t+\tau)}$$

$$= \overline{\int_{-\infty}^{\infty} h(\sigma_1)x(t-\sigma_1)d\sigma_1 \int_{-\infty}^{\infty} h(\sigma_2)x(t+\tau-\sigma_2)d\sigma_2} \tag{5.79}$$

By interchanging the order of integration and averaging, we then have

$$\phi_{yy}(\tau) = \int_{-\infty}^{\infty}\int_{-\infty}^{\infty} h(\sigma_1)h(\sigma_2)\overline{x(t-\sigma_1)x(t+\tau-\sigma_2)}d\sigma_1 d\sigma_2$$

$$= \int_{-\infty}^{\infty}\int_{-\infty}^{\infty} h(\sigma_1)h(\sigma_2)\phi_{xx}(\tau-\sigma_2+\sigma_1)d\sigma_1 d\sigma_2 \tag{5.80}$$

We now make the change of variable $\sigma_3 = \sigma_2 - \sigma_1$ to obtain

$$\phi_{yy}(\tau) = \int_{-\infty}^{\infty}\int_{-\infty}^{\infty} h(\sigma_1)h(\sigma_3+\sigma_1)\phi_{xx}(\tau-\sigma_3)d\sigma_1 d\sigma_3$$

$$= \int_{-\infty}^{\infty} \phi_{hh}(\sigma_3)\phi_{xx}(\tau-\sigma_3)d\sigma_3 \tag{5.81}$$

---

[10] M. Schetzen, "Fourier Transforms of Positive Functions" Mass. Inst. of Technology, Research Lab of Electronics Quarterly Progress Report No. 52, Jan 15, 1959, pp. 79–86.

in which $\phi_{hh}(\sigma)$, the autocorrelation function of $h(t)$, is defined as

$$\phi_{hh}(\sigma) = \int_{-\infty}^{\infty} h(\sigma_1)h(\sigma + \sigma_1)d\sigma_1 \tag{5.82}$$

Note from Eq. (5.36) that $\phi_{hh}(\sigma)$ is the convolution of $h(t)$ with $h(-t)$. Thus from Eq. (5.48), the Fourier transform of $\phi_{hh}(\sigma)$ equals the Fourier transform of $h(t)$ times the Fourier transform of $h(-t)$. Now, from Eq. (5.40), $H(j\omega)$ is the Fourier transform of $h(t)$ and with the use of Eq. (5.22) we have that the Fourier transform of $h(-t)$ is $H^*(j\omega)$. Thus

$$\begin{aligned} \Phi_{hh}(\omega) &= \int_{-\infty}^{\infty} \phi_{hh}(\sigma)e^{-j\omega\sigma}d\sigma \\ &= H(j\omega)H^*(j\omega) = |H(j\omega)|^2 \end{aligned} \tag{5.83}$$

We now observe from Eq. (5.81) that $\phi_{yy}(\tau)$ is the convolution of $\phi_{hh}(\tau)$ with $\phi_{xx}(\tau)$. That is, in terms of the notation of Eq. (5.36),

$$\phi_{yy}(\tau) = \phi_{hh}(\tau) * \phi_{xx}(\tau) \tag{5.84}$$

With the use of Eq. (5.48) we then have that the Fourier transform of $\phi_{yy}(\tau)$ is

$$\begin{aligned} \Phi_{yy}(\omega) &= \Phi_{hh}(\omega)\Phi_{xx}(\omega) \\ &= |H(j\omega)|^2\Phi_{xx}(\omega) \end{aligned} \tag{5.85}$$

in which, from Eqs. (5.44), $|H(j\omega)|$ is the gain of the LTI system. With the use of Eq. (5.84) we obtain by substituting Eq. (5.85) into Eq. (5.78b)

$$\overline{y^2(t)} = \phi_{yy}(0) = \int_{-\infty}^{\infty} \Phi_{yy}(\omega)d\omega \tag{5.86a}$$

$$= \int_{-\infty}^{\infty} |H(j\omega)|^2\Phi_{xx}(\omega)d\omega \tag{5.86b}$$

With this relation, we can show that $\Phi_{xx}(\omega)$ is the physical power density spectrum that one would measure in the laboratory. Our demonstration parallels our discussion of the energy density spectrum. To observe this let $y(t)$ be the voltage across the 1-ohm resistor in Fig. 5.3. Then the total average power dissipated in the resistor is $\phi_{yy}(0) = \overline{y^2(t)}$. Because the right side of Eq. (5.86a) is the integral overall frequencies of $\Phi_{yy}(\omega)$, we can interpret it as a power density spectrum in watts/radian per second. For this reason, $\Phi_{yy}(\omega)$ is called the *power density spectrum* of $y(t)$.

To make the interpretation of a power density spectrum more concrete, consider an ideal method to measure the power density spectrum of a waveform $x(t)$. Ideally, to measure the average power of $x(t)$ in a band of frequencies, we would apply $x(t)$ to an ideal low-pass filter with the cut-off frequency ω_1. An ideal low-pass filter is one that has unity gain in the passband and zero gain for all frequencies above the cut-off frequency so that the system gain $|H(j\omega)|$ is as defined by Eq. (5.55).

As shown in Fig. 5.3, the output of the ideal low-pass filter, Eq. (5.55), is a voltage that is applied to a 1-ohm resistor. Thus, with the use of Eq. (5.86b), the average power dissipated in the 1-ohm resistor is

$$\phi_{yy}(0) = \overline{y^2(t)} = \int_{-\omega_1}^{\omega_1} \Phi_{xx}(\omega)d\omega \tag{5.87}$$

Similarly, to measure the total power contained by the waveform $x(t)$ in the frequency band $\omega_1 < \omega < \omega_2$, we would use an ideal bandpass filter with unity gain in the given band and zero gain outside the given band. In accordance with Eq. (5.86b), the total power dissipated by the 1-ohm resistor is then

$$\phi_{yy}(0) = \overline{y^2(t)} = \int_{-\omega_2}^{-\omega_1} \Phi_{xx}(\omega)d\omega + \int_{\omega_1}^{\omega_2} \Phi_{xx}(\omega)d\omega \tag{5.88}$$

Because, as we discussed, the power density spectrum $\Phi_{xx}(\omega)$ is an even function of ω, both integrals on the right side of Eq. (5.88) have the same value so that we also can write

$$\phi_{yy}(0) = \overline{y^2(t)} = 2\int_{\omega_1}^{\omega_2} \Phi_{xx}(\omega)d\omega \tag{5.89}$$

Observe that the experimental measurement of the power contained by a real time function $x(t)$ in a given frequency band is numerically equal to twice the area under $\Phi_{xx}(\omega)$ in the given band. That is, $\Phi_{xx}(\omega)$ is the actual power density spectrum of $x(t)$ in watts per radian/second that we would ideally measure in the laboratory. Furthermore, observe from our discussion that $\Phi_{xx}(\omega) \geq 0$ and is a real function of ω.

In summary, the autocorrelation of a real time function is the time average given by Eq. (5.70). Wiener showed that the Fourier transform of the autocorrelation function exists and is the power density spectrum of the continuing waveform. This is the Wiener theorem of autocorrelation that is given by the Fourier transform pair, Eqs. (5.78). We then showed that the Fourier transform of the autocorrelation function is the actual power density spectrum of the real time function that one would experimentally measure in the laboratory. A problem with the theory we've discussed to this point is that we generally do not have a detailed knowledge of the continuing waveform and so we have no way of analytically calculating the autocorrelation function of such real time functions. We thus form a statistical model of the waveform from which the autocorrelation function can be determined analytically.

IV. Stochastic Processes

As we discussed, the power density spectrum $\Phi_{ff}(\omega)$ of a continuing waveform $f(t)$ is equal to the Fourier transform of $\phi_{ff}(\tau)$, the autocorrelation function of $f(t)$. The autocorrelation function is the time average

$$\phi_{ff}(\tau) = \overline{f(t)f(t+\tau)} \tag{5.90}$$

For theoretical analysis, we thus first determine the autocorrelation function of a waveform in order to determine its power density spectrum. A problem with determining the autocorrelation function is that the waveform $f(t)$ is rarely known in sufficient detail to determine the desired average analytically as an integration over time as given by Eq. (5.70). We thus require a subterfuge.

First note that the time average of a waveform is just dependent on the fraction of time that the waveform has certain values and is not dependent on the time order at which these values occur. For example, the time average of a rectangular waveform that jumps between the values zero and A is equal to A times the fraction of time that the value of the waveform is A irrespective of the times at which the waveform jumps between zero and A. As another example, the time average of $\sin(\omega t + \phi)$ is zero no matter what the value of ω or ϕ. Observe then that it is only necessary to know certain general characteristics of a waveform in order to determine its average. Specifically, we note that the time average of a waveform is unchanged if the order of the amplitudes of the waveform is rearranged. In consequence, if the members of a set of waveforms differ only in the arrangement of their amplitudes, then each waveform of the set will have the same time average. It is this observation that lies at the base of using stochastic process theory to determine the time average of a waveform.

In stochastic process theory, one imagines repeating an experiment an infinite number of times. The set of all the possible experimental outcomes is called an ensemble and each possible experimental outcome is a member of the ensemble. For example, consider a function $f(n)$ constructed by flipping a coin. The procedure for constructing the function $f(n)$ is to make $f(n) = 1$ if the coin lands heads on the nth toss and to make $f(n) = -1$ if the coin lands tails on the nth toss. The function so constructed is then a sequence of 1s and -1s. Thus, if the coin is considered to be an "honest" one, we would expect the coin to be a head for $1/2$ of the tosses and a tail for $1/2$ of the tosses. The expected average of $f(n)$ is thus zero even though the exact sequence of 1s and -1s of $f(n)$ is not known.

We now can create an ensemble in the following manner. We do a *gedanken* experiment (which is an experiment performed only in one's mind) by imagining the coin flipping experiment being performed in the same manner with similar coins by an infinite number of people. Each person's experiment results in a different function of n because the sequence of 1s and -1s obtained by each person is different. Each of the infinite number of functions obtained by our *gedanken* experiment is a member of the ensemble. Call the kth member function of this ensemble $f_k(n)$. Because the function is constructed in the same manner with a similar coin by each person, we expect that the members of the ensemble will differ from each other only in a rearrangement of its 1s and -1s. Thus, if each person used an honest coin, we would expect the average of each function, $f_k(n)$, to be zero.

Now consider each person's n^{th} toss; that is, we consider $f_k(n)$ for a given value of n. As we vary k by going down the ensemble for the given value of n, a sequence of values of $f_k(n)$ is obtained. Denote the sequence of 1s and -1s obtained for various values of k as η_n. The sequence η_n will be different for different values of n and is called the random variable obtained as one goes down the ensemble at n. The fraction of 1s of the sequence η_n is the probability of obtaining a head on the nth toss, which we denote by $P_{\eta_n}(1)$.[11] Similarly, $P_{\eta_n}(-1)$ is the probability of obtaining a tail on the nth toss. Denote the specific values that the random variable η_n can have by y_m. For our example, $y_m = \pm 1$. We then define $P_{\eta_n}(y_m)$ as the probability that $\eta_n = y_m$. If the manner in which each person tosses his coin doesn't change with n, we expect that the fraction of people who obtain a head doesn't change with n. For such a situation, $P_{\eta_n}(y_m)$ does not change with n. If that is so, we say the ensemble is first-order stationary. If the ensemble is first-order stationary, then $P_{\eta_n}(y_m)$ doesn't change with n so that we can drop the subscript n and write the probability as $P_\eta(y_m)$.

Now, if the fraction of 1s obtained as one goes down the ensemble doesn't change with n, then the fraction of 1s in the whole ensemble is equal to $P_\eta(1)$. Similarly, because each member function of the ensemble contains the same fraction of 1s, we conclude that the fraction of 1s in the whole ensemble is equal to the fraction of 1s in any member function. Consequently, because things equal to the same thing are equal to each other, we conclude that $P_\eta(1)$ is also equal to the fraction of 1s in any member function of the ensemble. We thus conclude that $P_\eta(y_m)$ is equal to the fraction of the sequence of any member function of the ensemble that has the value y_m. Such an ensemble is said to be first-order ergodic.[12]

That is, a first-order ergodic ensemble is one in which the following two quantities are equal:

1)   the fraction of time that any member function of the ensemble has a given value (called the time aggregate) and
2)   at any given time instant, the fraction of the member functions that has the given value (called the esemble aggregate).

We now can generalize our discussion by considering the construction of an ensemble of time functions in which the value of any member function is one of a given set of values, $y_1, y_2, y_3, \ldots$ and in which each member function of the ensemble has the same time aggregate. This means that the fraction of time any one member function has the value y_m (for $m = 1, 2, 3, \ldots$) is the same as that for any other member function. If such an ensemble is first-order stationary,

---

[11] This is often called the statistical definition of probability. There also is an axiomatic definition, which was developed by the eminent Russian mathematician Andrei N. Kolmogorov. The axiomatic approach is discussed in his text, *Foundations of the Theory of Probability*, Chelsea Publishing Co., second English edition, 1956. There are some problems in rigorous mathematics with statistical definition that the axiomatic definition avoids.

[12] In mathematics, ergodic theory is rather complex. In the mathematical view, the ensemble aggregate at $n = n_2$ is considered to be a mapping of the ensemble aggregate at $n = n_1$. The mathematical problem is to determine the mapping properties for which the ensemble is ergodic. For the ensemble to be ergodic, the mapping must be one that keeps the values well mixed over the ensemble members so that all the ensemble members have the same time aggregate and so are equal to the ensemble aggregate. In consequence, mathematicians sometimes refer to ergodic theory as the well-mixing theory. We have circumvented the mathematicians' concerns by our method of constructing an ensemble.

we then can conclude from our discussion that the ensemble is first-order ergodic because the time aggregates and the ensemble aggregates are then the same.

Because the time and ensemble aggregates of a first-order ergodic ensemble are the same, we observe that the average obtained going down the ensemble is equal to the time average of any one ensemble member. That is, the time and the ensemble averages are equal. We express this as $\overline{f(t)} = \bar{\eta}$ in which $f(t)$ is an ensemble member and η is the random variable of amplitude obtained by going down the ensemble at a time instant. The overbar indicates the average. That is, $\overline{f(t)}$ is the time average of the time function and $\bar{\eta}$ is the ensemble average of the random variable η.[13] Because $P_\eta(y_m)$ is the fraction of ensemble members whose value is y_m, we observe that

$$\overline{f(t)} = \bar{\eta} = \sum_m y_m P_\eta(y_m) \tag{5.91}$$

Also, observe that any function of a time aggregate will have the same average as the same function of an ensemble aggregate. Thus, if the function is the square of the values, we have for a first-order ergodic ensemble

$$\overline{f^2(t)} = \overline{\eta^2} = \sum_m y_m^2 P_\eta(y_m) \tag{5.92}$$

So far, we've considered ensembles for which the aggregates are composed of a set of discrete values. What if the member functions are continuous functions of time? In such a case, the ensemble cannot be ergodic. The problem lies in the number of amplitudes involved. Each member of our ensemble is obtained by performing a *gedanken* experiment. Thus ensemble member 1 is obtained from experiment 1, ensemble member 2 is obtained from experiment 2, and so on. We thus can assign an integer to each of the infinite number of ensemble members. This is the same as saying that the members of the ensemble can be counted. For this reason, such an infinity is called a countable infinity. However, all the values of any ensemble member that is a continuous function of time cannot be counted. This is not because it is a practical impossibility, but rather it is theoretically impossible! That is, we cannot even consider a *gedanken* experiment in which we assign an integer to each value of a continuous function. The values of a continuous function are thus said to be uncountable. This astonishing statement is easily shown by proving that the assumption that they are countable leads to an absurdity. For this let us just consider all the values between zero and one. Each value in this range can be expressed as a decimal. If they are countable, we can list them as follows:

First number $0.\alpha_1\alpha_2\alpha_3\alpha_4\ldots$
Second number $0.\beta_1\beta_2\beta_3\beta_4\ldots$

---

[13] The ensemble average $\bar{\eta}$ is also called the expectation of the random variable η and denoted by $E(\eta)$. The possible confusion between time and ensemble averages is avoided by using Roman letters for time functions and Greek letters for random variables.

Third number $0.\gamma_1 \gamma_2 \gamma_3 \gamma_4 \ldots$
Fourth number $0.\delta_1 \delta_2 \delta_3 \delta_4 \ldots$
etc.

In this listing, each of the letters is an integer. Our assumption is that every decimal value between zero and one is contained in this list. To show that this assumption is false, we'll choose a number that is not a member of this list. One such number is $0.abcd\ldots$ in which $a \neq \alpha_1$, $b \neq \beta_2$, $c \neq \gamma_3$, $d \neq \delta_4$, etc. That is, the number chosen differs from the first number on our list at least in the first decimal place, it differs from the second number on our list at least in the second decimal place, and generally, it differs from the nth number on our list at least in the nth decimal place. (To avoid any ambiguities that can arise because, for example, $0.2000\ldots = 0.1999\ldots$, we do not use an infinite sequence of 9s in the decimal number we choose.) If you argue that the chosen number is the only missing number, just add it to the list and construct a new number between zero and one that is not on the new list by using the same procedure. The list thus cannot contain all the numbers between zero and one, and so the set of values between zero and one is uncountable. Egodicity means that the time aggregates are the same as the ensemble aggregates. The ensemble of continuous time functions thus cannot be ergodic because the time aggregate of each time function has an uncountable number of values while the ensemble aggregate contains only a countable number of values because there are only a countable number of member functions of the ensemble.

The reason for our interest in egodicity is, as we showed in Eq. (5.91) and Eq. (5.92), that it is a sufficient condition for the equality of time and ensemble averages so that we can replace one with the other. There is, however, a subterfuge to relax the ergodic requirement for the equality of ensemble and time averages. The subterfuge is to break the interval of amplitude values into infinitesimal ranges and consider the fraction of time that $y \leq f(t) < y + dy$ for an ensemble member. Note that there are only a countable number of such infinitesimal ranges. This fraction of time will be the same for each ensemble member if the ensemble members of our constructed ensemble differ from one another only by a rearrangement of amplitudes. We thus call that fraction of time a time aggregate of $f(t)$.

Now consider the ensemble aggregate obtained by going down the ensemble at a given instant of time. The ensemble aggregate is then the fraction of ensemble members for which $y \leq \eta < y + dy$. Again note that there are only a countable number of such intervals. The ensemble aggregate will be the same at each instant of time if our constructed ensemble is first-order stationary. The fraction of ensemble members for which $y \leq \eta < y + dy$ is equal to $P_\eta(y)dy$ in which $P_\eta(y)$ is called the probability density of the random variable η. With this modified definition of aggregates, we have only a countable number of aggregate values. We thus can use our previous discussion to show that if the time aggregate of each function is the same and if the ensemble is first-order stationary, then the time and ensemble aggregates are equal. Such an ensemble is called first-order quasi-ergodic because only the time and ensemble aggregates of values grouped in dy intervals are equal. We are not concerned with the distribution of the values within a dy interval because that does not affect the averages.

Consequently, we also have that the time and ensemble averages are equal for quasi-ergodic ensembles. That is

$$\overline{f(t)} = \lim_{T \to \infty} \frac{1}{2T} \int_{-T}^{T} f(t)dt = \bar{\eta} = \int_{-\infty}^{\infty} y P_\eta(y)dy \qquad (5.93)$$

and any function of the aggregates will have the same average. Thus, if the function is the square of the values, we have for a first-order quasi-ergodic ensemble

$$\overline{f^2(t)} = \lim_{T \to \infty} \frac{1}{2T} \int_{-T}^{T} f^2(t)dt = \overline{\eta^2} = \int_{-\infty}^{\infty} y^2 P_\eta(y)dy \qquad (5.94)$$

To illustrate the concepts we've discussed so far, consider the ensemble of time functions $f_k(t) = \cos(\omega_0 t + z_k)$ in which the phase z_k varies down the ensemble with a probability density distribution $P_\zeta(z)$. That is, the fraction of ensemble members for which $z \le \zeta < z + dz$ is $P_\zeta(z)dz$. We thus have an ensemble of sinusoids, each with the frequency ω_0 and amplitude equal to one but with a phase angle z that varies from ensemble member to ensemble member. For our example, we choose all phase angles to be equally likely so that

$$P_\zeta(z) = \begin{cases} \dfrac{1}{2\pi} & \text{for} \quad 0 \le z < 2\pi \\ 0 & \text{otherwise} \end{cases} \qquad (5.95)$$

With this choice, all angles of the sinusoid are equally likely to be obtained as we go down the ensemble at any given time instant. The ensemble thus is first order stationary. Also, every member function has the same time aggregate because, except for a phase angle z, each is the same function. The ensemble thus is quasi-ergodic. The ensemble aggregate at any instant of time is thus equal to the time aggregate of any one ensemble member. The ensemble aggregate is $P_\eta(y)dy$ in which $P_\eta(y)$ is the amplitude probability density. Consequently, the probability density $P_\eta(y)$ can be determined by determining the time aggregate of any one member function of the ensemble as follows: Choose the kth member of the ensemble and, for convenience, let $\theta = \omega_0 t + z_k$.

Now note from Fig. 5.4 that $P_\eta(y)dy$, which equals the percentage of time that $y < \cos(\theta) < y + dy$, is

$$\frac{2|d\theta|}{2\pi} = \frac{|d\theta|}{\pi} \qquad (5.96)$$

From the relation $y = \cos(\theta)$ we have

$$dy = -\sin\theta \, d\theta = -\sqrt{1 - \cos^2\theta} \, d\theta = -\sqrt{1 - y^2} \, d\theta \qquad (5.97)$$

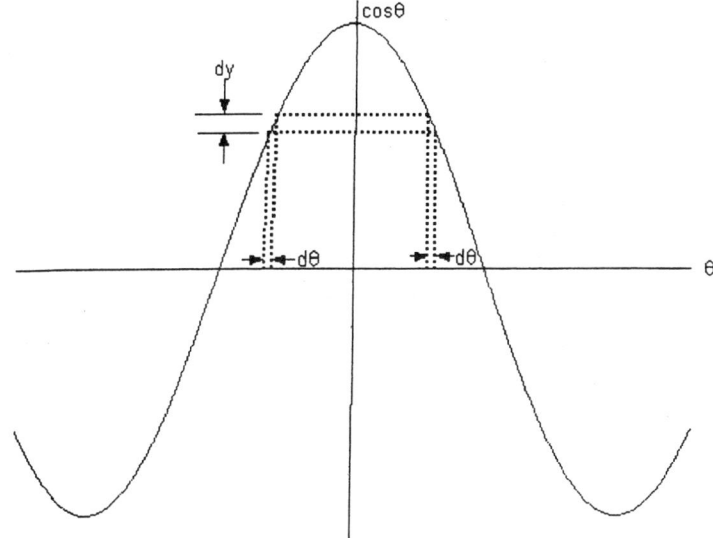

Fig. 5.4   The calculation of the amplitude density of cos θ.

so that

$$|d\theta| = \frac{dy}{\sqrt{1 - y^2}} \tag{5.98}$$

and consequently,

$$P_\eta(y) = \begin{cases} \dfrac{1}{\pi\sqrt{1 - y^2}} & \text{for} \quad |y| < 1 \\ 0 & \text{for} \quad |y| > 1 \end{cases} \tag{5.99}$$

Note that the average $\bar{\eta}$ is

$$\bar{\eta} = \int_{-\infty}^{\infty} y P_\eta(y)\,dy = \int_{-1}^{1} \frac{y}{\pi\sqrt{1 - y^2}}\,dy = 0 \tag{5.100}$$

as expected from Eq. (5.93) because the time average of any ensemble member is zero. Also, the average of η^2 is

$$\overline{\eta^2} = \int_{-\infty}^{\infty} y^2 P_\eta(y)\,dy = \int_{-1}^{1} \frac{y^2}{\pi\sqrt{1 - y^2}}\,dy$$

$$= \frac{2}{\pi}\left[\frac{y\sqrt{1 - y^2}}{2} + \frac{1}{2}\sin^{-1}(y) \right]_0^1 = \frac{2}{\pi}\frac{1}{2}\sin^{-1}(1) = \frac{1}{2} \tag{5.101a}$$

as expected from Eq. (5.94) because this is equal to the time average

$$\overline{\sin^2(\omega_0 t + \phi)} = \frac{1}{2} - \frac{1}{2}\overline{\cos(2\omega_0 t + 2\phi)} = \frac{1}{2} - 0 = \frac{1}{2} \qquad (5.101b)$$

All the concepts developed now can be extended to two dimensions in the following manner. We again construct an ensemble in which each member function is the result of a *gedanken* experiment as discussed previously. Let η_1 be the random variable of amplitude obtained by going down the ensemble at a time instant $t = t_1$, and let η_2 be the random variable of amplitude obtained by going down the ensemble at the time instant $t = t_2$. Following our discussion, we define the second-order probability density $P\eta_1\eta_2\,(y_1,\ y_2;\ t_1,\ t_2)$. It is defined by $P\eta_1\eta_2(y_1,\ y_2;\ t_1,\ t_2)\,dy_1\,dy_2$ being equal to the fraction of ensemble members for which $y_1 \le \eta_1 < y_1 + dy_1$ at $t = t_1$ and also $y_2 \le \eta_2 < y_2 + dy_2$ at $t = t_2$. The ensemble is said to be second-order stationary if the second-order probability density does not depend on what point of the time axis is called $t = 0$. For second-order stationary ensembles, the second-order probability density thus only depends on the time difference $\tau = t_2 - t_1$ and we write the second-order probability density of a second-order stationary ensemble as $P\eta_1\eta_2(y_1, y_2; \tau)$.

We now define the second-order time aggregate as the fraction of time that the amplitude of an ensemble member is between y_1 and $y_1 + dy_1$ and its amplitude τ seconds later is between y_2 and $y_2 + dy_2$. If each ensemble member is obtained by imagining the same experiment being performed a countably infinite number of times, we will have that the second-order time aggregate of each ensemble member will be the same. Furthermore, if the ensemble is second-order stationary, then by arguing in the same manner as previously, we conclude that the second-order time aggregates will be the same as the second-order ensemble aggregates. An ensemble for which this is so is said to be second-order quasi-ergodic.

Because the second-order time and ensemble aggregates are the same, the average of any function of their values will be the same. Thus, for a second-order quasi-ergodic ensemble, we have that the ensemble average

$$\overline{\eta_1\eta_2} = \int_{-\infty}^{\infty} \int_{-\infty}^{\infty} y_1 y_2 P\eta_1\eta_2(y_1, y_2; \tau)\,dy_1 dy_2 \qquad (5.102)$$

is equal to the time average

$$\phi_{ff}(\tau) = \overline{f(t)f(t+\tau)} = \lim_{T\to\infty} \frac{1}{2T}\int_{-T}^{T} f(t)f(t+\tau)\,dt \qquad (5.103)$$

The function $\phi_{ff}(\tau)$ is the autocorrelation function of $f(t)$ defined by Eq. (5.70).

Often, the autocorrelation function cannot be determined by calculating the time average in Eq. (5.103) because the waveform $f(t)$ is not known in sufficient detail. In such cases, by constructing a second-order quasi-ergodic ensemble in the *gedanken* manner we've discussed, the autocorrelation function can be determined as

$$\phi_{ff}(\tau) = \overline{\eta_1 \eta_2} \tag{5.104}$$

which is the ensemble average given by Eq. (5.102). The computation of the ensemble average only requires knowledge of certain statistics that often can be determined from the manner by which the ensemble is constructed.

Because the received waveform of an airborne Doppler radar is not known in sufficient detail, we shall determine its power density spectrum by first constructing a second-order stationary ensemble of Doppler radar received waveforms by means of a *gedanken* experiment as discussed in this section. Equation (5.104) will then be used to determine the autocorrelation function. The power density spectrum of the Doppler radar received waveform will then be determined using Eq. (5.78a), which is the Fourier transform of the autocorrelation function.

Chapter 6

Thin Antenna Pattern

THE ANALYSIS of the Doppler echo in the general case involves a number of variables that tend to obscure a physical interpretation of our final results. We'll thus begin by analyzing a simplified case and interpret the results obtained before considering the general case. For this, we'll begin by analyzing the case in which the antenna is directed along the flight path and the antenna pattern is thin; that is, its width in the azimuth direction is very narrow. This simplification eliminates the azimuth coordinate in our analysis and so eliminates a number of parameters involved in our analysis, such as the effect of the antenna pattern width in a direction perpendicular to the flight path and pointing the antenna in a direction off the flight path. The thin antenna case thus will enable us to obtain analytic approximations from which a physical interpretation of the contribution of the various parameters to the Doppler spectrum can be obtained. This understanding will assist us in the interpretation of the expressions we'll obtain for the general case in which the expressions are involved and require a computer for their evaluation.

I.   The Echo, $e_r(t)$

For our analysis, consider the schematic shown in Fig. 6.1. The antenna is a distance h above the average level of the terrain and is traveling parallel to it in the x direction at a constant velocity v. The z axis is perpendicular to the average level of the terrain, which is at $z = 0$. The origin of the coordinate is fixed relative to the terrain so that at a time t, the antenna is above the point $d = vt$ on the x axis. We consider an antenna that illuminates a thin strip of constant width in the y direction and is centered along the x axis as shown in Fig. 6.1.[1]

Let the illuminated strip shown in Fig. 6.1 be subdivided into incremental areas of length dx and width equal to the full illuminated width. Consider an incremental area at the point x. The ratio of the amplitude of the wave reflected back to the antenna from the incremental area to the amplitude of the incident wave is $\rho(x, \psi)dx$, in which the backscattering coefficient $\rho(x, \psi)$ is a function of the position x of the incremental area and of the incidence angle ψ. We

---

[1] We shall use rectangular coordinates to obtain the desired theoretical results. Later, we'll express our theoretical results in polar coordinates because that is the coordinate system in which antenna patterns are described.

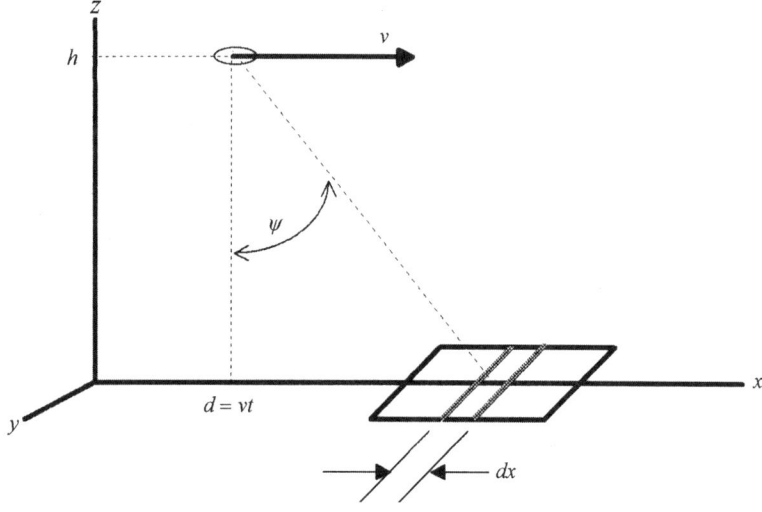

Fig. 6.1   Geometric representation for the thin antenna pattern.

consider the continuous wave (CW) case in which the signal radiated from the antenna is[2]

$$e(t) = E_0 \cos(\omega_0 t + \theta). \tag{6.1}$$

The infinitesimal waveform at the antenna terminals received from the incremental area is then

$$de_r(t) = E_0 \rho(x, \psi) s(x - d) \cos\left[\omega_0 t + \theta - \frac{4\pi r}{\lambda} - \beta(x, \psi)\right] dx \tag{6.2}$$

In this expression, $s(x - d)$ is the two-way antenna pattern as projected on the terrain. It includes such propagation effects as the inverse square law. The wavelength is λ and the phase delay angle $4\pi r/\lambda$ is the sinusoidal phase shift due to the propagation delay time incurred by the wave traveling the distance $2r = 2[h^2 + (x - d)^2]^{1/2}$. The other phase term $\beta(x, \psi)$ is the phase shift of the reflection arising from two sources: the dielectric properties of the terrain in the dx region and the roughness of the terrain surface that results in the path length deviating from $2r$.

---

[2] Observe that the relation between the radiated waveform and the received waveform is linear. Consequently, other cases such as pulse Doppler radar can be analyzed by expanding the pulsed waveform in a Fourier series and using superposition with the results we obtain for the CW case.

The echo is the sum of the incremental backscattered waves so that

$$e_r(t) = \int_{-\infty}^{\infty} E_0 \rho(x, \psi) \, s(x - d) \cos[\omega_0 t + \theta - 4\pi r/\lambda - \beta(x, \psi)]dx \qquad (6.3)$$

in which $r^2 = h^2 + (x - d)^2$ and $d = vt$.

II. Autocorrelation Function of the Echo

As discussed in section 5.3, we first obtain the autocorrelation function of the echo $\phi_{rr}(\tau)$ from which the power density spectrum of the echo $\Phi_{rr}(\omega)$ can be determined. In accordance with Eq. (5.70), the autocorrelation function of the echo is

$$\phi_{rr}(\tau) = \lim_{T \to \infty} \frac{1}{2T} \int_{-\infty}^{\infty} e_r(t) e_r(t + \tau) dt \qquad (6.4)$$

Then, in accordance with the Wiener theorem, Eq. (5.78a),

$$\Phi_{rr}(\omega) = \frac{1}{2\pi} \int_{-\infty}^{\infty} \phi_{rr}(\tau) e^{-j\omega\tau} \, d\tau \qquad (6.5)$$

As described in section 5.4, the autocorrelation function will be determined by constructing a quasi-ergodic ensemble of time functions for which the ensemble average is equal to the desired time average $\phi_{rr}(\tau)$. The desired ensemble is constructed by letting $\theta, \rho(x, \psi)$, and $\beta(x, \psi)$ be random variables down the ensemble such that

1) θ is uniformly distributed between 0 and 2π,
2) θ is independent of $\rho(x, \psi)$ and $\beta(x, \psi)$, and
3) $\rho(x, \psi)$ and $\beta(x, \psi)$ are second-order stationary random processes of the argument x.

Then

$$\phi_{rr}(\tau) = \overline{e_r(t) e_r(t + \tau)} \qquad (6.6)$$

in which the bar indicates the ensemble average.

The properties of random variables can be interpreted physically by noting a method for constructing the ensemble. Each member of the ensemble is obtained by a *gedanken* experiment. In each experiment, we denote some point on the x axis as the origin ($x = 0$) and observe the return $e_r(t)$ for a given phase of the transmitter. The choice of $x = 0$ is random and is independent of the transmitter phase θ, which is chosen so that all phase angles are equally likely. Each member of the ensemble then corresponds to the return $e_r(t)$ from a different *gedanken* experiment.

Our calculations are simplified by using the identity $2 \cos\phi = e^{j\phi} + e^{-j\phi}$ to express $e_r(t)$ in complex form as

$$2\,e_r(t) = e_c(t) + e_c^*(t) \tag{6.7}$$

in which the star indicates the conjugate. From Eq. (6.3), the complex echo $e_c(t)$ is

$$e_c(t) = \int_{-\infty}^{\infty} E_0\rho(x,\,\psi)\,s(x - d)\,e^{j[\omega_0 t + \theta - (4\pi r/\lambda) - \beta(x,\,\psi)]}\,dx \tag{6.8}$$

in terms of $e_c(t)$, the desired autocorrelation function is

$$
\begin{aligned}
4\phi_{rr}(\tau) &= \overline{[e_c(t) + e_c^*(t)][e_c(t + \tau) + e_c^*(t + \tau)]} \\
&= \overline{e_c(t)e_c^*(t + \tau)} + \overline{e_c^*(t)e_c(t + \tau)}
\end{aligned}
\tag{6.9}
$$

because

$$\overline{e_c(t)e_c(t + \tau)} = 0. \tag{6.10}$$

Equation (6.10) can be verified by noting that $\overline{e_c(t)e_c(t + \tau)}$ will contain the multiplicative factor $\overline{e^{j2\theta}}$, which is zero because θ is uniformly distributed over all angles and is independent of $\rho(x,\,\psi)$ and $\beta(x,\,\psi)$.

We now define

$$\phi_{cc}(\tau) = \overline{e_c^*(t)e_c(t + \tau)} \tag{6.11}$$

Because $z + z^* = 2\,\mathrm{Re}\{z\}$, we then have from Eq. (6.9) that

$$\phi_{rr}(\tau) = \frac{1}{2}\mathrm{Re}\{\phi_{cc}(\tau)\} \tag{6.12}$$

By use of Eq. (6.12) and from section 5.3 $\phi_{cc}(\tau) = \phi_{cc}^*(-\tau)$, we have that the power density spectrum of the echo is

$$\Phi_{rr}(\omega) = \frac{1}{2}[\Phi_{cc}(\omega) + \Phi_{cc}(-\omega)] \tag{6.13}$$

in which

$$\Phi_{cc}(\omega) = \frac{1}{2\pi}\int_{-\infty}^{\infty} \phi_{cc}(\tau)e^{-j\omega\tau}\,d\tau \tag{6.14}$$

is the spectrum of the complex echo $e_c(t)$. From Eq. (6.8), the autocorrelation function $\phi_{cc}(\tau)$ given by Eq. (6.11) is

$$
\begin{aligned}
\phi_{cc}(\tau) = E_0^2\,e^{j\omega_0\tau}\int_{-\infty}^{\infty}\int_{-\infty}^{\infty} &\overline{\rho(x_1,\psi_1)\,\rho(x_2,\psi_2)\,s(x_1 - vt)\,s(x_2 - vt - v\tau)\cdots} \\
&\overline{e^{-j(4\pi/\lambda)(r_2 - r_1)}e^{-j[\beta(x_2,\psi_2) - \beta(x_1,\psi_1)]}}\,dx_1\,dx_2
\end{aligned}
\tag{6.15}
$$

The reflection coefficient has a certain mean value determined by the character of the terrain on which is superimposed a random component due to terrain irregularities. We thus express the backscatter coefficient $\rho(x, \psi)$ in the form

$$\rho(x, \psi) = b_S(\psi)\rho_r(x, \psi) \tag{6.16}$$

in which

$$b_S(\psi) = \overline{\rho(x, \psi)} \tag{6.17}$$

For a terrain that is a diffuse surface, $b_S(\psi)$ corresponds to the Lambert law. The factor $b_S(\psi)$ is not a function of x because $\rho(x, \psi)$ is a stationary random variable. For a constant altitude h we define the quantity $b(x - d)$ to be

$$b(x - d) = b(h \tan \psi) = b_S(\psi) \tag{6.18}$$

We then have from Eq. (6.16) and Eq. (6.17) that $\overline{\rho_r(x, \psi)} = 1$ independent of the angle ψ.

To simplify the derivation of the Doppler spectrum, we now combine similar type terms. Later, we'll analyze the contribution of each component to the Doppler spectrum. For this, we first define the complex antenna pattern $p(x - d)$ to be

$$p(x - d) = s(x - d)e^{-j(4\pi/\lambda)r} \tag{6.19}$$

in which, from Fig 6.1, $r = [h^2 + (x - d)^2]^{1/2}$. Also, we define the effective complex antenna pattern $g(x - d)$ as

$$g(x - d) = b(x - d)p(x - d) \tag{6.20}$$

This is the complex antenna pattern $p(x - d)$ modified by $b(x - d)$, the expected value of the backscatter coefficient. By substituting these defined quantities in Eq. (6.15), we have

$$\phi_{cc}(\tau) = E_0^2 \, e^{j\omega_0 \tau} \int_{-\infty}^{\infty} \int_{-\infty}^{\infty} g^*(x_1 - vt)g(x_2 - vt - v\tau) \cdots$$

$$\overline{\rho_r(x_1, \psi_1)\rho_r(x_2, \psi_2)e^{j[\beta(x_2, \psi_2) - \beta(x_1, \psi_1)]}} \, dx_1 \, dx_2 \tag{6.21}$$

Equation (6.21) is the general expression for the complex autocorrelation function $\phi_{cc}(\tau)$.

III.   Autocorrelation of the Reflection Coefficient

The specific evaluation of the integral in Eq. (6.21) depends on knowledge of the correlation function

$$\phi_{x\psi}(x_1, x_2, \psi_1, \psi_2) = \overline{\rho_r(x_1, \psi_1)\rho_r(x_2, \psi_2)e^{j[\beta(x_2, \psi_2) - \beta(x_1, \psi_1)]}} \qquad (6.22)$$

For constant altitude h and velocity v, we observe from Fig. 6.1 that

$$\left. \begin{array}{l} z_1 = h \tan \psi_1 = x_1 - vt \\ z_2 = h \tan \psi_2 = x_2 - v(t - \tau) \end{array} \right\} \qquad (6.23)$$

Thus, the correlation function $\phi_{x\psi}$ in Eq. (6.22) can be expressed in the form

$$\phi_{x\psi}(x_1, x_2, \psi_1, \psi_2) = \phi_{xz}(x_1, x_2, z_1, z_2,)$$
$$= \phi(x_2, x_1, z_2 - z_1, z_1) \qquad (6.24)$$

The function ϕ is not a function of x_1 or x_2 independently but of $x_2 - x_1$ because the ensemble is stationary.

To obtain more specific expressions for the spectrum, we need to consider various specific cases. We'll analyze four cases that correspond to different possible physical situations. In all four cases we'll consider that the variation of ϕ in Eq. (6.24) with respect to only z_1 is negligible over the beam. The function ϕ then can be expressed as

$$\phi(x_2 - x_1, z_2 - z_1, z_1) = \phi(x_2 - x_1, z_2 - z_1, z_0) \qquad (6.25)$$

in which $z_0 = h \tan \psi_0$ and ψ_0 is the angle to the beam center. The four cases are:

1) $\phi(x_2 - x_1, z_2 - z_1, z_1) = c_1(z_0)\phi_1(x_2 - x_1)$
2) $\phi(x_2 - x_1, z_2 - z_1, z_1) = c_2(z_0)\phi_2(x_2 - x_1)\delta(z_2 - z_1)$
3) $\phi(x_2 - x_1, z_2 - z_1, z_1) = c_3(z_0)\phi_3(z_2 - z_1)$
4) $\phi(x_2 - x_1, z_2 - z_1, z_1) = c_4(z_0)\delta(x_2 - x_1)\phi_4(z_2 - z_1)$

in which $\delta(\cdot)$ is the unit impulse.[3]

In case 1, the correlation of the reflections from two given positions are independent of their angular difference for $|\psi_2 - \psi_1|$ less than a beamwidth. Opposing this, case 2 states that the reflections from two given positions within the beam are independent for a small angular difference. In case 3, the correlation of the reflection from two positions within the beam is independent of their separation for a constant angular difference. Finally, contrasting with case 3, case 4 states that the reflections from two positions are independent for small separations.

---

[3] As we discussed in section 5.2, please note that the width of an impulse is infinitesimal but not zero.

The four cases can be interpreted in terms of a model in which the surface is composed of many individual scatterers. For case 1, the scatterers can be considered to be spheres. For case 2, the scatterers are very rough compared with a wavelength so that reflections at two different angles are independent. For case 3, the terrain is smooth. The model for case 4 is that of rough terrain in which the reflections from different scatterers are independent.

Observe that for each of the four cases, ϕ is expressed as the product of a function of distance $(x_2 - x_1)$ and a function of angle $(z_2 - z_1)$. The correlation function ϕ for each of the four cases thus can be expressed in the form

$$\phi(x_2 - x_1, z_2 - z_1, z_1) = \phi_d(x_2 - x_1)\phi_\psi(z_2 - z_1) \tag{6.26}$$

The function $\phi_\psi(z_2 - z_1)$ includes $c(z_0)$, which is simply a constant because we consider the variation of ϕ in Eq. (6.24) with respect to only z_1 is negligible over the beam. Because $c(z_0)$ is a constant, it affects only the magnitude but not the shape of $\phi_{cc}(\tau)$.

IV. Power Density Spectrum of the Echo

The autocorrelation function $\phi_{cc}(\tau)$ obtained by substituting Eq. (6.26) in Eq. (6.21) is

$$\phi_{cc}(\tau) = E_0^2 e^{j\omega_0\tau} \int_{-\infty}^{\infty} \int_{-\infty}^{\infty} \phi_d(x_2 - x_1)\phi_\psi(x_2 - x_1 - v\tau)\ldots$$

$$g^*(x_1 - vt)g(x_2 - vt - v\tau)dx_1 dx_2. \tag{6.27}$$

With the change in variables $x_3 = x_1 - vt$ and $x_4 = x_2 - x_1$ in Eq. (6.27) and integrating with respect to x_3, we obtain

$$\phi_{cc}(\tau) = E_0^2 e^{j\omega_0\tau} \int_{-\infty}^{\infty} \int_{-\infty}^{\infty} \phi_d(x_4)\phi_a(x_4 - v\tau)dx_4 \tag{6.28}$$

in which

$$\phi_a(x) = \phi_\psi(x)\phi_g(x) \tag{6.29}$$

and $\phi_g(\tau)$ is the autocorrelation function of the complex antenna pattern $g(x)$ given by Eq. (6.20). That is

$$\phi_g(\tau) = \int_{-\infty}^{\infty} g^*(x)g(x + \tau)dx \tag{6.30a}$$

from which the Fourier transform of $\phi_g(\tau)$ is

$$\Phi_g(\omega) = |G(\omega)|^2 \tag{6.30b}$$

in which

$$G(\omega) = \frac{1}{2\pi} \int_{-\infty}^{\infty} g(x) e^{-j\omega x} dx \tag{6.30c}$$

is the Fourier transform of $g(x)$.

By substituting Eq. (6.28) in Eq. (6.14) we now can obtain the desired expression for the power density spectrum of the complex echo

$$\Phi_{cc}(\omega) = \frac{1}{2\pi} E_0^2 \int_{-\infty}^{\infty} e^{-j(\omega-\omega_0)\tau} d\tau \int_{-\infty}^{\infty} \phi_d(x_4)\phi_a(x_4 - v\tau) dx_4 \tag{6.31}$$

Equation (6.31) can be integrated by multiplying and dividing by $e^{-j[(\omega-\omega_0)/v]x_4}$ and letting $x_5 = x_4 - v\tau$

$$\Phi_{cc}(\omega) = \frac{1}{2\pi v} E_0^2 \int_{-\infty}^{\infty} \phi_d(x_4) e^{-j[(\omega-\omega_0)/v]x_4} dx_4 \int_{-\infty}^{\infty} \phi_a(x_5) e^{+j[(\omega-\omega_0)/v]x_5} dx_5$$

$$= \frac{2\pi}{v} E_0^2 \Phi_d\left(\frac{\omega - \omega_0}{v}\right) \Phi_a\left(-\frac{\omega - \omega_0}{v}\right) \tag{6.32}$$

in which

$$\Phi_d(\omega) = \frac{1}{2\pi} \int_{-\infty}^{\infty} \phi_d(x) e^{-j\omega x} dx \tag{6.33}$$

is the spectrum of the position-dependent component of the correlation function ϕ. The other spectrum factor is

$$\Phi_a(\omega) = \frac{1}{2\pi} \int_{-\infty}^{\infty} \phi_a(x) e^{-j\omega x} dx \tag{6.34}$$

By virtue of Eq. (6.29) and Eqs. (6.30), we have from a property of Fourier transforms that this spectrum can be expressed as the convolution

$$\Phi_a(\omega) = \int_{-\infty}^{\infty} |G(\sigma)|^2 \Phi_\psi(\omega - \sigma) d\sigma \tag{6.35}$$

In this expression,

$$G(\omega) = \frac{1}{2\pi} \int_{-\infty}^{\infty} g(x) e^{-j\omega x} dx \tag{6.36}$$

is the Fourier transform of the effective complex antenna pattern, while

$$\Phi_\psi(\omega) = \frac{1}{2\pi} \int_{-\infty}^{\infty} \phi_\psi(x) e^{-j\omega x} \, dx \tag{6.37}$$

is the spectrum arising from the angular-dependent component of the correlation function ϕ. Thus $\Phi_\alpha(\omega)$ is the spectrum due to the angular dependence of the echo. From Eq. (6.32), the spectrum of the complex echo is observed to be the product of the two spectra: $\Phi_d(\omega)$, which arises from variations of the return with position, and $\Phi_\alpha(\omega)$, which arises from variations of the return with angle.

For convenience in interpreting Eq. (6.32), we define $\omega' = \omega - \omega_0$. Then,

$$\Phi_{cc}(\omega_0 + \omega') = \frac{2\pi}{v} E_0^2 \Phi_d\left(\frac{\omega'}{v}\right) \Phi_\alpha\left(-\frac{\omega'}{v}\right) \tag{6.38}$$

To study the echo spectrum, some special cases will be examined. We first consider the factor $\Phi_d(\omega'/v)$ and then the factor $\Phi_\alpha(-\omega'/v)$, which, as we shall see, gives rise to the Doppler effect.

V.   Doppler Spectrum for Smooth and Periodic Terrain

Consider the case of smooth terrain corresponding to case 3 in section 6.3. Such a condition could exist over a smooth water surface or for small values of ω_0 for which the wavelength λ is large compared with the irregularities of the terrain. For such a case, $\phi_d(x) = \gamma_0^2$, so that

$$\Phi_d(\omega) = \gamma_0^2 \delta(\omega) \tag{6.39}$$

in which $\delta(\omega)$ is the unit impulse. Substituting this result in Eq. (6.38), we obtain

$$\Phi_{cc}(\omega_0 + \omega') = \frac{2\pi}{v} E_0^2 \gamma_0^2 \delta\left(\frac{\omega'}{v}\right) \Phi_\alpha\left(-\frac{\omega'}{v}\right)$$

$$= 2\pi E_0^2 \gamma_0^2 \Phi_\alpha(0) \delta(\omega') \tag{6.40}$$

The echo spectrum for this case is an impulse at $\omega = \omega_0$ so that the echo is a sinusoid of the same frequency as that of the transmitter. Thus, there is no Doppler effect for a condition satisfying case 3, a result sometimes referred to as "the smooth earth paradox".

A more general case for which there is no Doppler effect is that of periodic terrain, for which $\phi_d(x)$ is periodic with a fundamental period of T_x. With the use of Fourier series analysis, the expression for $\Phi_d(\omega)$, the Fourier transform of $\phi_d(x)$, can be expressed as

$$\Phi_d(\omega) = \sum_{n=-\infty}^{\infty} A_n \delta(\omega + n\omega_x) \tag{6.41}$$

in which $\omega_x = 2\pi/T_x$. Such a condition could be approximated by a vehicle flying over water with a periodic surface structure. Substituting Eq. (6.41) in Eq. (6.38) gives

$$\Phi_{cc}(\omega_0 + \omega') = \frac{2\pi}{v} E_0^2 \Phi_a\left(-\frac{\omega'}{v}\right) \sum_{n=-\infty}^{\infty} A_n \delta\left(\frac{\omega'}{v} + n\omega_x\right)$$

$$= 2\pi E_0^2 \sum_{n=-\infty}^{\infty} A_n \Phi_a(n\omega_x)\delta(\omega' + nv\omega_x) \qquad (6.42)$$

We observe from this expression that the echo spectrum is a series of impulses at the angular frequencies $\omega = \omega_0 + nv\omega_x$, $n = 0, \pm 1, \pm 2, \ldots$ so that the echo is a sinusoid of the same frequency as that of the transmitter, which is amplitude modulated by a periodic waveform with a fundamental angular frequency $\omega_m = v\omega_x$ for which the fundamental period of the modulating waveform is $T_m = (1/v)T_x$. Thus, there is no Doppler effect for this case and the velocity v can be determined from the echo spectrum only if the fundamental period of the terrain, T_x, is known. The results for smooth terrain can be obtained by letting $T_x \to \infty$.

VI. Doppler Spectrum for Rough Terrain: General Expressions

We now consider the echo spectrum for the usual physical case of rough terrain, which corresponds to case 4 in section 6.3. For such terrain, $\phi_d(x)$ is a narrow pulse. Thus from Fourier transform theory, $\Phi_d(\omega)$ is broad relative to $\Phi_a(\omega)$ so that from Eq. (6.38)

$$\Phi_{cc}(\omega_0 + \omega') = \frac{2\pi}{v} E_0^2 \Phi_d(0)\Phi_a\left(-\frac{\omega'}{v}\right) \qquad (6.43)$$

In practice, the width of $\Phi_a(-\omega'/v)$ is small in comparison with ω_0 so that from Eq. (6.13), the echo spectrum for rough terrain is

$$\Phi_{rr}(\omega_0 + \omega') = \frac{\pi}{v} E_0^2 \Phi_d(0)\Phi_a\left(-\frac{\omega'}{v}\right) \qquad (6.44)$$

To interpret this equation for the echo spectrum, we begin by analyzing $\Phi_a(\omega)$.

As given by Eq. (6.35), the spectrum $\Phi_a(\omega)$ is the convolution of $\Phi_\psi(\omega)$ with $|G(\omega)|^2$. To study the spectrum, we first shall consider the term $\Phi_\psi(\omega)$. We then shall study the other term, $|G(\omega)|^2$, from which, we shall see, the Doppler effect is obtained.

For our discussion of $\Phi_\psi(\omega)$, there are two cases of practical importance: one in which $\phi_\psi(x)$ is a narrow pulse compared with $\phi_g(x)$ and one in which $\phi_\psi(x)$ is a broad pulse compared with $\phi_g(x)$. Physically, $\phi_\psi(x)$ being a narrow pulse corresponds to the case for which the backscattering coefficient is a sensitive function of angle so that the reflections from two slightly different angles within the antenna beam are essentially uncorrelated. For rough terrain, such a condition could occur at large values of the frequency ω_0 for which the wavelength is small compared with the irregularities of the scatterers. For such a case, we observe from Eq. (6.29) that $\phi_a(x)$ also will be a narrow pulse so that its Fourier transform, $\Phi_a(\omega)$, is broad. We conclude then from Eq. (6.44) that for such a condition the power density spectrum of the echo will be equally broad, making it difficult to determine the presence of any Doppler effect. Because the wavelength of an optical laser is very small, Doppler radars using optical lasers can be expected to encounter this difficulty over certain terrains.

If $\phi_\psi(x)$ is a pulse that is broad in comparison with $\phi_g(x)$, we have from Eq. (6.29) that a good approximation of $\phi_a(x)$ is

$$\phi_a(x) = \phi_\psi(0)\phi_g(x) \tag{6.45}$$

Physically, $\phi_\psi(x)$ being a broad pulse corresponds to the case for which the backscattering coefficient is not a sensitive function of angle for the range of angles within the beam. By use of Eq. (6.35) we then can express $\Phi_a(\omega)$ as

$$\Phi_a(\omega) = 2\pi\phi_\psi(0)|G(\omega)|^2 \tag{6.46}$$

in which $G(\omega)$ is given by Eq. (6.36). We then have from Eq. (6.44) that

$$\Phi_{rr}(\omega_0 + \omega') = \frac{2\pi^2}{v} E_0^2 \phi_\psi(0)\phi_d(0) \left| G\left(-\frac{\omega'}{v}\right) \right|^2 \tag{6.47}$$

We note from Eq. (6.47) that for rough terrain in which the backscatter is not a very sensitive function of angle for the range of angles within the beam, the echo's power density spectrum is equal to a constant times that of $|G(-\omega'/v)|^2$ moved up in frequency by ω_0 because we defined $\omega' = \omega - \omega_0$. It is this term from which the Doppler effect is obtained.

To study the echo spectrum using Eq. (6.47), we have from Eq. (6.20) and Eq. (6.36) that the effective complex antenna pattern transform is the Fourier transform of the product of $b(x)$ and $p(x)$

$$G(-\omega) = \frac{1}{2\pi} \int_{-\infty}^{\infty} b(x)p(x)e^{j\omega x}\, dx \tag{6.48}$$

so that $G(-\omega)$ is also given by the convolution

$$G(-\omega) = \int_{-\infty}^{\infty} B(\sigma)P(\omega - \sigma)d\sigma \tag{6.49}$$

in which

$$P(\omega) = \frac{1}{2\pi} \int_{-\infty}^{\infty} p(x)e^{j\omega x}\, dx \tag{6.50}$$

and

$$B(\omega) = \frac{1}{2\pi} \int_{-\infty}^{\infty} b(x)e^{j\omega x} \, dx \qquad (6.51)$$

For our analysis of the spectrum, we shall determine $P(\omega)$ and $B(\omega)$ separately and then convolve them in accordance with Eq. (6.49) to obtain $G(\omega)$. Note in Eq. (6.47) that the quantity of interest is $P(\omega/v)$. With the use of Eq. (6.19) and Eq. (6.50), we have

$$P\left(\frac{\omega}{v}\right) = \frac{1}{2\pi} \int_{-\infty}^{\infty} p(x)e^{j(\omega/v)x}dx$$

$$= \frac{1}{2\pi} \int_{-\infty}^{\infty} s(x)e^{-j[(4\pi r/\lambda)-(\omega x/v)]} \, dx \qquad (6.52)$$

in which $r^2 = h^2 + x^2$. We shall interpret this expression by considering some special cases of practical interest. We shall ignore the inverse-square law for analytic simplicity in our interpretations

VII. Doppler Spectrum for a Narrow Antenna Pattern

We first consider the case of a narrow antenna pattern centered at $x = x_0$ as shown in Fig. 6.2. We define a narrow antenna pattern as one for which $s(x) \ll s(0)$ for $|x > x_0| > \delta$ in which $\delta \ll x_0$. We then can approximate the

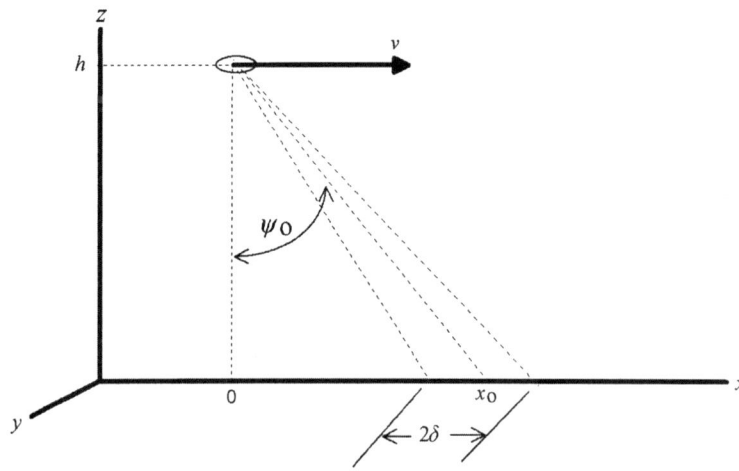

Fig. 6.2 Geometric representation for the narrow antenna pattern.

range r in Eq. (6.52) as

$$
\begin{aligned}
r &= [h^2 + x^2]^{1/2} \\
&= [h^2 + (x - x_0 + x_0)^2]^{1/2} \\
&= [h^2 + x_0^2 + 2x_0(x - x_0) + (x - x_0)^2] \\
&\approx [r_0^2 + 2x_0(x - x_0)]^{1/2} = r_0\left[1 + 2\frac{x_0(x - x_0)}{r_0^2}\right]^{1/2} \\
&\approx r_0\left[1 + \frac{x_0(x - x_0)}{r_0^2}\right] = r_0\left[\frac{h^2 + x x_0}{r_0^2}\right] = \frac{h^2 + x x_0}{r_0}
\end{aligned}
\tag{6.53}
$$

in which $r_0^2 = h^2 + x_0^2$. We now observe from Fig. 6.2 that $h/r_0 = \cos \psi_0$ and $x_0/r_0 = \sin \psi_0$ so that the above expression for r can be written as

$$
r \approx h \cos \psi_0 + x \sin \psi_0
\tag{6.54}
$$

Using this approximation for r in Eq. (6.52), the pattern transform for this case is

$$
\begin{aligned}
P\left(\frac{\omega}{v}\right) &= e^{-j(4\pi/\lambda)h\cos\psi_0}\frac{1}{2\pi}\int_{-\infty}^{\infty} s(x)e^{j[(\omega/v)-(4\pi/\lambda)\sin\psi_0]x}\,dx \\
&= e^{-j(4\pi/\lambda)h\cos\psi_0}\,S\left(\frac{\omega}{v} - \frac{4\pi}{\lambda}\sin\psi_0\right)
\end{aligned}
\tag{6.55}
$$

in which

$$
S(\omega) = \frac{1}{2\pi}\int_{-\infty}^{\infty} s(x)e^{j\omega x}\,dx
\tag{6.56}
$$

is the Fourier transform of the two-way antenna pattern. We now consider the effect of the terrain reflectivity on the spectrum.

A.   The Spectrum for the Terrain Parameter $b(x) = b_0$

For the special case in which $b(x)$ is a constant equal to b_0 over the range $|x - x_0| < \delta$, we have from Eq. (6.48) and Eq. (6.55)

$$
\left|G\left(-\frac{\omega}{v}\right)\right|^2 = b_0^2\left|S\left(\frac{\omega}{v} - \frac{4\pi}{\lambda}\sin\psi_0\right)\right|^2
\tag{6.57}
$$

so that from Eq. (6.47), the echo power density spectrum is

$$
\Phi_{rr}(\omega_0 + \omega') = \frac{2\pi^2}{v}E_0^2 b_0^2 \phi_\psi(0)\phi_d(0)\left|S\left(\frac{\omega'}{v} - \frac{4\pi}{\lambda}\sin\psi_0\right)\right|^2
\tag{6.58}
$$

We observe for this case that the echo spectrum has the shape of the transform of the two-way antenna pattern centered about the angular frequency $\omega = \omega_0 + \omega_d$ in which

$$\omega_d = \frac{4\pi v}{\lambda} \sin \psi_0 \quad \text{rad/s} \qquad (6.59a)$$

or

$$f_d = \frac{\omega_d}{2\pi} = \frac{2v}{\lambda} \sin \psi_0 \quad \text{hertz} \qquad (6.59b)$$

is the Doppler shift.

B.   The Effect of the Terrain Parameter $b(x)$ on the Echo Spectrum

For the narrow beam case, the variation of $b(x)$ over the beam can be approximated by a linear variation with x so that it can approximated over the beam as

$$b(x) = b_0 - m(x - x_0) \qquad (6.60)$$

We then have from Eq. (6.20) that

$$\begin{aligned} g(x) &= [b_0 - m(x - x_0)]\,p(x) \\ &= (b_0 + mx_0)p(x) - mxp(x) \end{aligned} \qquad (6.61)$$

To obtain the transform of this equation, note from Eq. (6.50) that

$$-j\frac{d}{d\omega}P(\omega) = \frac{1}{2\pi}\int_{-\infty}^{\infty} xp(x)e^{j\omega x}\,dx \qquad (6.62)$$

With this relation, the transform of Eq. (6.61) is

$$G(-\omega) = (b_0 + mx_0)P(\omega) + jm\frac{d}{d\omega}P(\omega) \qquad (6.63)$$

The desired expression for $|G(-\omega/v)|^2$ is obtained by substituting Eq. (6.55) in Eq. (6.63).

To illustrate the application of this result, consider the case for which the two-way antenna pattern $s(x)$ is symmetric about $x = x_0$ so that

$$s(x) = s_0(x - x_0) \qquad (6.64)$$

in which $s_0(x)$ is an even function of x. We then have from Eq. (6.56) that

$$S(\omega) = S_0(\omega)e^{j\omega x_0} \qquad (6.65)$$

in which $S_0(\omega)$ is the Fourier transform of $s_0(x)$. Thus

$$S_0(\omega) = \frac{1}{2\pi} \int_{-\infty}^{\infty} s_0(x)e^{j\omega x}\,dx \tag{6.66}$$

is a real and even function of ω because $s_0(x)$ is a real and even function of x. Substituting Eq. (6.65) and Eq. (6.55) in Eq. (6.63), we then obtain

$$G\left(-\frac{\omega}{v}\right) = \left[b_0 S_0\left(\frac{\omega}{v} - \frac{4\pi}{\lambda}\sin\psi_0\right)\right.$$
$$\left. + jmS_0'\left(\frac{\omega}{v} - \frac{4\pi}{\lambda}\sin\psi_0\right)\right]e^{j((\omega x_o/v)-(4\pi/\lambda)h\cos\psi_0)} \tag{6.67}$$

so that

$$\left|G\left(-\frac{\omega}{v}\right)\right|^2 = b_0^2\left|S_0\left(\frac{\omega}{v} - \frac{4\pi}{\lambda}\sin\psi_0\right)\right|^2 + m^2\left|S_0'\left(\frac{\omega}{v} - \frac{4\pi}{\lambda}\sin\psi_0\right)\right|^2 \tag{6.68}$$

To interpret this result, consider Fig. 6.3 in which ω_d is given by Eq. (6.59a). As shown, $\left|S_0((\omega - \omega_d)/v)\right|^2$ is a positive and even function about $\omega = \omega_0 + \omega_d$ because $S_0(\omega)$ is, for our example, a real and even function of ω. Thus $\left|S_0'((\omega - \omega_0)/v)\right|^2$ is, as shown, a positive and even function that is zero at $\omega = \omega_0 + \omega_d$. The sum $\left|G(-(\omega/v))\right|^2$ thus has the form shown. We thus observe that for the narrow beam case in which $s(x)$ is symmetric about $x = x_0$, the effect of the terrain parameter $b(x)$ is to broaden the echo power

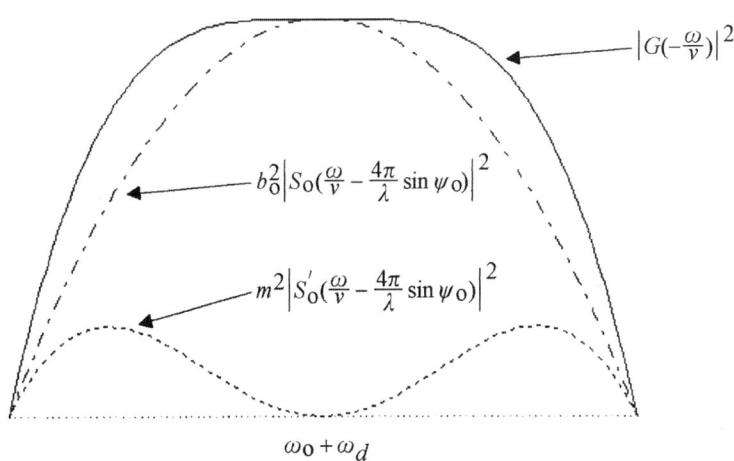

Fig. 6.3 Spectrum broadening due to terrain backscatter for a narrow antenna pattern.

density spectrum. The center of the spectrum, however, is not altered so that no bias is introduced in the experimental determination of the Doppler frequency.

VIII.   Doppler Spectrum for a Gaussian Antenna Pattern

A better understanding of the spectrum can be obtained by analyzing an analytic approximation for a wider antenna pattern than the narrow pattern discussed above. This can be obtained for the special case of a Gaussian antenna pattern for which

$$s(x) = \frac{1}{a\sqrt{2\pi}} e^{-((x-x_0)^2/2a^2)} \tag{6.69}$$

In order not to detract from the basic analysis of the results, many of the basic calculations for this case are included in the appendices.

A.   The Spectrum for $b(x) = b_0$

For an antenna pattern given by Eq. (6.69) and for the special case in which the terrain parameter $b(x)$ is a constant equal to b_0, it is shown in Appendix A that

$$\left| G\left(-\frac{\omega}{v}\right) \right|^2 = \frac{b_0^2}{2\pi\sqrt{1 + (4\pi a^2/\lambda r_0)^2}} e^{-(\omega-\omega_d)^2/2\sigma^2} \tag{6.70}$$

in which

$$\omega_d = \frac{4\pi v}{\lambda} \sin \psi_0 \text{ rad/s} \tag{6.71a}$$

is the Doppler shift and the bandwidth is determined by

$$2\sigma^2 = \left(\frac{v}{a}\right)^2 \left[1 + \left(\frac{4\pi a^2}{\lambda r_0}\right)^2 \right] \text{ rad/s} \tag{6.71b}$$

We observe that the case of a Gaussian ground illumination pattern for which $b(x)$ is a constant, the power density spectrum of the echo also is Gaussian in shape centered at $\omega = \omega_0 + \omega_d$ in which ω_d is the usual Doppler shift given by Eq. (6.59a). The spectrum bandwidth given by Eq. (6.71b) is σ. In terms of σ, the half-power bandwidth is 2.355σ.

The accuracy in the experimental measurement of the Doppler frequency is determined by how large the Doppler spectrum bandwidth σ is relative to the Doppler shift ω_d. The smaller the ratio σ to ω_d, the greater the accuracy. From Eqs. (6.71), this meaningful parameter is:

$$2\left(\frac{\sigma}{\omega_d}\right)^2 = \left(\frac{\lambda}{4\pi a \sin \psi_0}\right)^2 \left[1 + \left(\frac{4\pi a^2}{\lambda r_0}\right)^2 \right] \tag{6.72}$$

Equation (6.72) can be expressed in terms of the antenna angular beamwidth α by substituting the relations

$$\frac{a}{r_0} = \frac{\alpha}{2\cos\psi_0} \quad \text{and} \quad \frac{h}{r_0} = \cos\psi_0 \tag{6.73}$$

obtained from Fig. A.1 in Appendix A. The result is

$$2\left(\frac{\sigma}{\omega_d}\right)^2 = \left(\frac{\alpha}{\sin 2\psi_0}\right)^2\left[1 + \left(\frac{\lambda\cos^3\psi_0}{\pi h\alpha^2}\right)^2\right] \tag{6.74}$$

Equation (6.74) is a minimum for $\alpha = \alpha_{min}$ in which

$$\alpha_{min} = \left[\frac{\lambda}{\pi h}\cos^3\psi_0\right]^{1/2} \text{ radians} \tag{6.75}$$

This angle is small. For example, for a wavelength $\lambda = 3$ cm, a height $h = 1$ kilometer, and a pointing angle $\psi_0 = 45$ degrees, the angle is $\alpha_{min} = 0.1$ degree. Substituting Eq. (6.75) in Eq. (6.74) we then obtain

$$\left.\frac{\sigma}{\omega_d}\right|_{min} = \left[\frac{\lambda}{4\pi h}\frac{\cos\psi_0}{\sin^2\psi_0}\right]^{1/2} \tag{6.76}$$

To better analyze Eq. (6.74), we express it in dimensionless form with the use of Eq. (6.75) as

$$2\left(\frac{\sigma}{\omega_d}\frac{\sin 2\psi_0}{\alpha_{min}}\right)^2 = \left(\frac{\alpha}{\alpha_{min}}\right)^2 + \left(\frac{\alpha_{min}}{\alpha}\right)^2 \tag{6.77}$$

Figure 6.4 is a graph of Eq. (6.77). The shape of the curve can be explained in terms of the model in which the surface is composed of many individual scatterers. For such a model, the echo from an individual scatterer as it passes through the antenna pattern is a pulse. The total echo is thus the sum of independent pulses because, in the present case, the reflections from different scatterers are independent. Also, because the shape of each pulse is the same in our present case, we have that the shape of the echo power density spectrum is the same as that of the energy density spectrum for an individual pulse.

Now, a pulse is a sinusoid at the transmitter frequency ω_0, which has been phase and amplitude modulated. The phase modulation is the result of the changing path length between the airborne radar and the scatterer. The amplitude modulation is a consequence of the antenna pattern and the terrain parameter $b(x)$. For a very small antenna beamwidth in which $\alpha \ll \alpha_{min}$, the frequency of each pulse can be considered to be constant equal to $\omega_0 + \omega_d$. The shape of the spectrum for this range of α is then the square of the magnitude of the Fourier transform of the pulse envelope. This is seen to correspond to the case of a

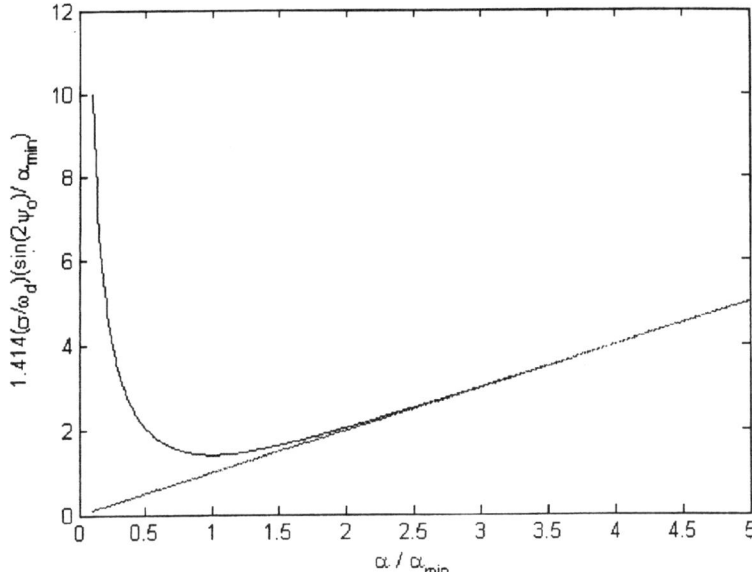

Fig. 6.4   Dimensionless graph of the spectrum bandwidth vs beamwidth.

narrow beam antenna. Increasing the antenna beamwidth α results in an increase of the pulse length and so results in a decrease in the echo spectrum bandwidth σ. However, as α approaches α_{min}, it is no longer a valid approximation to consider the pulse frequency to be constant. The changing phase modulation over the pulse length causes the spectrum to broaden for $\alpha > \alpha_{min}$. In accordance with this model, the spectrum $\Phi_{rr}(\omega)$ is seen to have the same shape as that of a chirp pulse. This suggests that an experimental method to study the spectrum for the case of a thin antenna is to generate such a pulse in the laboratory and measure its spectrum with a spectrum analyzer. The effect of the antenna pattern $s(x)$ and the terrain parameter $b(x)$ can then be evaluated by modifying the amplitude modulation of the chirp pulse in accordance with $s(x)b(x)$.

Observe from Fig. 6.4 and our discussion that both the bandwidth σ and the antenna beamwidth α cannot be small simultaneously so that there is a minimum value of the product $\sigma\alpha$. The velocity v is proportional to the Doppler frequency ω_d and so the uncertainty in the velocity Δv is proportional to the uncertainty in determining ω_d. The uncertainty in determining ω_d is proportional to spectrum width σ. Thus the uncertainty in velocity Δv is proportional to the spectrum width σ. Furthermore, the uncertainty of position on the terrain Δx is proportional to the antenna beamwidth α. Thus there is a minimum value of the product $(\Delta v)(\Delta x)$. The value of the product for the Gaussian antenna pattern is close to the lowest possible value. I call this an *Airborne Radar Doppler Uncertainty Principle*, which is similar to the Heisenberg Uncertainty Principle. As a consequence of this Doppler uncertainty principle, the validity of the determination of the echo spectrum using a model in which a frequency is assigned to each point on the terrain, as in the quasi-static approximation, is limited. Examples of this are illustrated in section 9.3 in our analysis of the general antenna pattern.

For Doppler radars operating in the microwave frequency range, the usual antenna beamwidth satisfies the condition $\alpha > 2\alpha_{min}$. From Fig. 6.4, we observe that for this beamwidth range, a good approximation of Eq. (6.77) is

$$\frac{\sigma}{\omega_d} = \frac{\alpha}{\sqrt{2}\sin 2\psi_0} \tag{6.78}$$

We define the Doppler quality factor Q_d as

$$Q_d = \frac{\omega_d}{2\sigma} \tag{6.79}$$

Note that Q_d is a measure of the velocity to the velocity uncertainty so that the larger Q_d, the more accurate the determination of the velocity. For $\alpha > 2\alpha_{min}$, we then have from Eq. (6.78) that

$$Q_d = \frac{1}{\alpha\sqrt{2}}\sin 2\psi_0 \tag{6.80}$$

Note from this result that at $\psi_0 = 45°$, Q_d is a maximum with a value equal to $(1/\alpha\sqrt{2})$ with α in radians or $(1/81\alpha)$ with α in degrees.

B. The Effect of the Terrain Parameter $b(x)$

For the narrow antenna pattern that is symmetric about x_0, we saw that to a first order the center of the echo spectrum was not altered. For the Gaussian pattern, we can obtain a slightly better first-order approximation of the effect of $b(x)$ upon the echo spectrum. This approximation is derived in Appendix B. The result obtained is a shift of the center of the spectrum. The approximate value of the shift δ of the spectrum center from the Doppler frequency ω_d obtained is

$$\frac{\delta}{\omega_d} \approx \frac{ma^2}{x_0 b_0} \tag{6.81}$$

This shift can be expressed in terms of the antenna beamwidth α by use of the relations obtained from Fig. A.1 in Appendix A

$$\frac{a}{r_0} = \frac{\alpha}{2\cos\psi_0}, \quad \frac{h}{r_0} = \cos\psi_0, \quad \text{and} \quad \frac{x_0}{h} = \tan\psi_0 \tag{6.82}$$

and

$$\frac{ma}{b_0} = -\frac{\alpha}{2}\frac{d}{d\psi}\ln b_S(\psi)\bigg|_{\psi=\psi_0} \tag{6.83}$$

in which $b_S(\psi) = b(h\tan\psi)$ as given by Eq. (6.18). The result of the substitution is

$$\frac{\delta}{\omega_d} \approx \frac{\alpha^2}{2\sin 2\psi_0}\left[-\frac{d}{d\psi}\ln b_S(\psi)\right]_{\psi=\psi_0} \tag{6.84}$$

Observe that the first-order approximation of the shift of the center frequency from the Doppler frequency is proportional to the square of the antenna beamwidth so that it is negligible for narrow antenna patterns as we found in section 6.7B.

We can relate the change of the center frequency to an effective change in the forward-look angle ψ_0. From the relation for the Doppler frequency, Eq. (6.71a), $f_d = (2v/\lambda)\sin\psi_0$, we have

$$\frac{\Delta f_d}{f_d} = -\frac{\delta}{\omega_d} = \frac{\Delta\psi_0}{\tan\psi_0} \tag{6.85}$$

so that

$$\Delta\psi_0 = -\frac{\delta}{\omega_d}\tan\psi_0 \tag{6.86}$$

Thus, by substituting Eq. (6.84) we obtain

$$\Delta\psi_0 \approx \left(\frac{\alpha}{2\cos\psi_0}\right)^2\left[\frac{d}{d\psi}\ln b_S(\psi)\right]_{\psi=\psi_0} \tag{6.87}$$

This is a first-order approximation of the effective change in the forward-look angle due to the terrain parameter $b(x)$ for a thin Gaussian antenna pattern as given by Eq. (6.69). For a narrow antenna pattern, note that to a first order, the shift in the forward-look angle is proportional only to the square of the angular beamwidth.

Chapter 7

General Antenna Pattern

T HE CASE of a thin antenna pattern pointed in the direction of travel was analyzed in the last chapter. Because the antenna was considered thin and
pointed in the direction of travel, the ground geometry was just one dimensional.
As a result, the thin antenna pattern greatly simplified our analysis because the
reflection from each particle as it travels through the beam has the same phase
modulation due to its changing path length to the antenna and the same amplitude
modulation due to the antenna pattern. This enabled us to determine effective
analysis techniques and to obtain a great deal of insight concerning the spectrum
of the echo. In this chapter, we analyze the general case in which the antenna is
not necessarily thin or directed in the direction of travel. Thus the amplitude
modulation and phase modulation of the reflection from different particles will
be different. The ground geometry in our general case thus is two dimensional
so that the equations we'll obtain are more complex. However, we shall follow
the analysis techniques we used in Chapter 6 for the simpler one-dimensional
case as a guide for our analysis of the general case. We then shall interpret our
results concerning the spectrum of the echo using, as a guide, the insight obtained
in our analysis of the case for a thin antenna pattern. Our development thus will
parallel that used in the last chapter.

I.   The Echo, $e_r(t)$

For our analysis, consider the schematic shown in Fig. 7.1. The conditions are
the same as those for Fig. 6.1 except, as shown, the antenna pattern is not necessarily centered along the x axis and is not necessarily very narrow. For our present
development, the shape of the antenna pattern on the ground is some general
two-dimensional shape.

For our analysis, let the illuminated area on the ground be subdivided into
incremental areas of length dx and width dy. Consider an incremental area centered at the point (x, y). The ratio of the amplitude of the wave reflected back to
the antenna from the incremental area to the amplitude of the incident wave is
$\rho(x, y, \psi) \, dxdy$ in which the backscattering coefficient $\rho(x, y, \psi)$ is a function of
the absolute position of the incremental area and of the angle of incidence ψ.
As in the previous chapter, the wave transmitted from the antenna is the sinusoid.

$$e(t) = E_0 \cos(\omega_0 t + \theta) \tag{7.1}$$

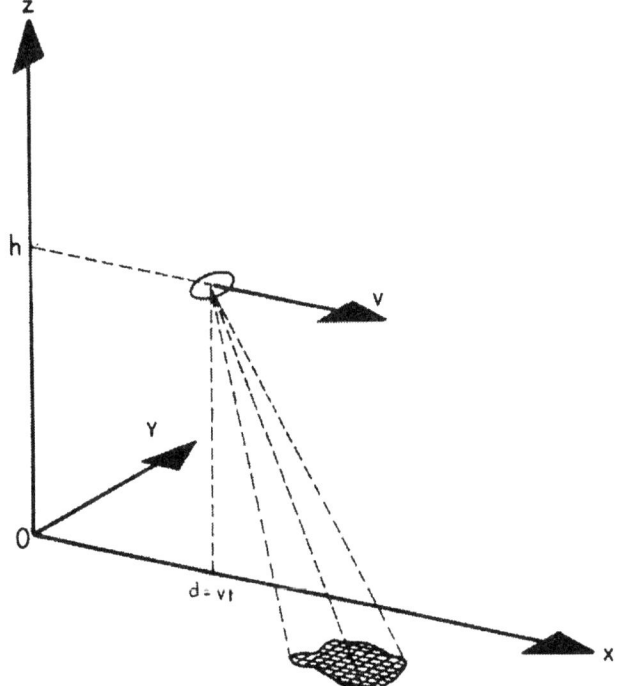

Fig. 7.1   Geometric representation for the general antenna pattern.

The infinitesimal received waveform that is backscattered from the incremental area is then

$$de_r(t) = E_0 \rho(x, \ y, \ \psi) \, s(x - d, \ y)$$

$$\times \cos\left[\omega_0 t + \theta - \frac{4\pi r}{\lambda} - \beta(x, \ y, \ \psi) \right] dxdy \qquad (7.2)$$

In this expression, $d = vt$, $s(x - d, y)$ is the two-way antenna pattern as projected on the terrain. It includes such propagation effects as the inverse-square law. The phase delay angle $4\pi r/\lambda$, in which λ is the wavelength, is the sinusoidal phase shift due to the propagation delay time incurred by the wave traveling the distance $2r = 2[h^2 + (x - d)^2 + y^2]^{1/2}$. The other phase term, $\beta(x, y, \psi)$, is the phase shift of the reflection arising from two sources: the dielectric properties of the terrain in the incremental area and the roughness of the terrain surface that results in the path length deviating from $2r$. The received echo is the sum of the incremental backscattered waves so that

$$e_r(t) = \int_{-\infty}^{\infty} \int_{-\infty}^{\infty} E_0 \rho(x, y, \psi) s(x - d, y)$$

$$\times \cos\left[\omega_0 t + \theta - \frac{4\pi r}{\lambda} - \beta(x, y, \psi)\right] dx\, dy \qquad (7.3)$$

II. Autocorrelation Function of the Echo

To determine the power density spectrum $\Phi_{rr}(\omega)$ of the received echo $e_r(t)$, we shall determine the autocorrelation function $\phi_{rr}(\tau)$ given by Eq. (6.4). As in section 6.2, we form a quasi-ergodic ensemble of time functions so that the ensemble average is equal to the desired time average $\phi_{rr}(\tau)$. As in Chapter 6, the desired ensemble is constructed by letting θ, $\rho(x, y, \psi)$, and $\beta(x, y, \psi)$ be random variables down the ensemble such that

1) θ is uniformly distributed between 0 and 2π,
2) θ is independent of $\rho(x, y, \psi)$ and $\beta(x, y, \psi)$, and
3) $\rho(x, y, \psi)$ and $\beta(x, y, \psi)$ are second-order stationary random variables.

Then, as in Chapter 6,

$$\phi_{rr}(\tau) = \overline{e_r(t)e_r(t + \tau)} \qquad (7.4)$$

in which the overbar indicates the ensemble average. Our expressions will be simplified by expressing the received echo $e_r(t)$ in the complex form given by Eq. (6.7) in which, from Eq. (7.3),

$$e_c(t) = \int_{-\infty}^{\infty} \int_{-\infty}^{\infty} E_0 \rho(x, y, \psi) s(x - d, y) e^{j[\omega_0 t + \theta - (4\pi r/\lambda) - \beta(x, y, \psi)]}\, dx\, dy \qquad (7.5)$$

In terms of $e_c(t)$, the desired autocorrelation function $\phi_{rr}(\tau)$ is given by Eq. (6.9) and the power density spectrum of the received echo $\Phi_{rr}(\omega)$ is given by Eq. (6.13).

With the use of Eq. (7.5), the autocorrelation function $\phi_{cc}(\tau)$ given by Eq. (6.11) is

$$\phi_{cc}(\tau) = E_0^2 e^{j\omega_0 \tau} \int_{-\infty}^{\infty} \int_{-\infty}^{\infty} \int_{-\infty}^{\infty} \int_{-\infty}^{\infty} \overline{\rho(x_1, y_1, \psi_1)\rho(x_2, y_2, \psi_2) s(x_1 - vt, y_1) \cdots}$$

$$\overline{s(x_2 - vt - v\tau, y_2) e^{j(4\pi/\lambda)[r_2 - r_1] - \beta(x, y, \psi)} e^{j[\beta(x_2, y_2, \psi_2) - \beta(x_1, y_1, \psi_1)]}}\, dx_1\, dx_2\, dy_1\, dy_2 \qquad (7.6)$$

Following Chapter 6, we express $\rho(x, y, \psi)$ in the form

$$\rho(x, y, \psi) = b_S(y, \psi)\rho_r(x, y, \psi) \qquad (7.7)$$

in which

$$b_S(y, \psi) = \overline{\rho(x, y, \psi)} \tag{7.8}$$

The average $b_S(y, \psi)$ has the same interpretation as in Chapter 6 so that for diffuse surfaces it would correspond to the Lambert law. The factor $b_S(y, \psi)$ is not a function of x because $\rho(x, y, \psi)$ is a stationary random variable. For a constant altitude h we shall define the quantity $b(x, y)$ to be

$$b(x - d, y) = b_S(y, \psi) \tag{7.9}$$

For the other factor in Eq. (7.7), we then have $\overline{\rho_r(x, y, \psi)} = 1$ independent of its argument.

Also, paralleling our definitions of Eq. (6.19) and Eq. (6.20), we define the complex antenna pattern $p(x, y)$ as

$$p(x, y) = s(x, y)e^{-j(4\pi/\lambda)} = s(x, y)e^{-j(4\pi/\lambda)\sqrt{h^2 + x^2 + y^2}} \tag{7.10}$$

and the effective complex antenna pattern $g(x, y)$ as

$$g(x, y) = b(x, y)p(x, y) \tag{7.11}$$

In terms of these defined quantities, Eq. (7.6) can be written as

$$\phi_{cc}(\tau) = E_0^2 e^{j\omega_0 \tau} \int_{-\infty}^{\infty} \int_{-\infty}^{\infty} \int_{-\infty}^{\infty} \int_{-\infty}^{\infty} g^*(x_1 - vt, y_1)g(x_2 - vt - v\tau, y_2) \cdots$$

$$\overline{\rho_r(x_1, y_1, \psi_1)\rho_r(x_2, y_2, \psi_2)e^{j[\beta(x_2, y_2, \psi_2) - \beta(x_1, y_1, \psi_1)]}} \, dx_1 dx_2 dy_1 dy_2 \tag{7.12}$$

Equation (7.12) is the general expression for the autocorrelation function $\phi_{cc}(\tau)$.

III.   Autocorrelation of the Reflection Coefficient

The specific evaluation of the integral in Eq. (7.12) depends upon knowledge of the correlation function

$$\phi_{x, y, \psi}(x_1, x_2, y_1, y_2, \psi_1, \psi_2) = \overline{\rho_r(x_1, y_1, \psi_1)\rho_r(x_2, y_2, \psi_2)}$$

$$\overline{\times e^{j[\beta(x_2, y_2, \psi_2) - \beta(x_1, y_1, \psi_1)]}} \tag{7.13}$$

For constant altitude h and velocity v let

$$z_1 = x_1 - vt \quad \text{and} \quad z_2 = x_2 - v[t + \tau] \tag{7.14}$$

Observe from Fig. 7.1 that the angle ψ is specified by z and y so that the correlation function in Eq. (7.13) can be expressed as

$$\phi_{x,y,\psi}(x_1, x_2, y_1, y_2, \psi_1, \psi_2) = \phi_{x,y,z}(x_1, x_2, y_1, y_2, z_1, z_2)$$

$$= \phi(x_2 - x_1, y_2 - y_1, z_2 - z_1, y_1, z_1) \qquad (7.15)$$

As in Eq. (6.24), this expression is not a function of x_1 or x_2 independently but of $x_2 - x_1$ because the ensemble is stationary.

Also, as in Eq. (6.25), we consider the case in which the variation of ϕ in Eq. (7.15) over the antenna beam as function of y_1 only or z_1 only is negligible so that we can express ϕ as

$$\phi(x_2 - x_1, y_2 - y_1, z_2 - z_1, y_1, z_1) = \phi(x_2 - x_1, y_2 - y_1, z_2 - z_1, y_0, z_0) \qquad (7.16)$$

in which (y_0, z_0) are the coordinates of the antenna beam center.

Parallel to our development in Chapter 6, we confine our analysis to the cases in which the terrain is smooth, periodic, and rough. For smooth terrain,

$$\phi(x_2 - x_1, y_2 - y_1, z_2 - z_1, y_0, z_0) = C_1(y_0, z_0)\phi_1(y_2 - y_1, z_2 - z_1) \qquad (7.17a)$$

For periodic terrain,

$$\phi(x_2 - x_1, y_2 - y_1, z_2 - z_1, y_0, z_0)$$

$$= C_2(y_0, z_0)\phi_2(x_2 - x_1, y_2 - y_1)\phi_3(y_2 - y_1, z_2 - z_1) \qquad (7.17b)$$

in which ϕ_2 is a periodic function of its arguments. Finally, for rough terrain,

$$\phi(x_2 - x_1, y_2 - y_1, z_2 - z_1, y_0, z_0)$$

$$= C_3(y_0, z_0)\delta(x_2 - x_1, y_2 - y_1)\phi_4(y_2 - y_1, z_2 - z_1) \qquad (7.17c)$$

The function ϕ in each case is expressed as the product of a function of distance $(x_2 - x_1, y_2 - y_1)$ and a function of angle $(y_2 - y_1, z_2 - z_1)$. We thus consider all these cases simultaneously by expressing ϕ in the form

$$\phi(x_2 - x_1, y_2 - y_1, z_2 - z_1, y_0, z_0)$$

$$= C(x_0, y_0)\phi_d(x_2 - x_1, y_2 - y_1)\phi_\psi(y_2 - y_1, z_2 - z_1) \qquad (7.18)$$

Note in this expression that the function ϕ_d depends only on position while the function ϕ_ψ depends only on angle.

IV. Power Density Spectrum of the Echo

The power density spectrum of the complex echo is the Fourier transform of the echo autocorrelation function $\phi_{cc}(\tau)$. First, the autocorrelation function of the complex echo is obtained by substituting Eq. (7.18) into Eq. (7.12) to obtain

$$\phi_{cc}(\tau) = E_0^2 C(x_0, y_0) e^{j\omega_0 \tau} \int_{-\infty}^{\infty} \int_{-\infty}^{\infty} \int_{-\infty}^{\infty} \int_{-\infty}^{\infty} \phi_d(x_2 - x_1, y_2 - y_1)$$

$$\times \phi_\psi(y_2 - y_1, x_2 - x_1 - v\tau) g^*(x_1 - vt, y_1) g(x_2 - vt - v\tau, y_2)$$

$$\times dx_1 \, dx_2 \, dy_1 \, dy_2 \tag{7.19}$$

To simplify this expression, we first make the change of variables $x_3 = x_1 - vt$, $x_4 = x_2 - x_1$, and $y_4 = y_2 - y_1$. Then, by integrating with respect to x_3 and y_1, we obtain

$$\phi_{cc}(\tau) = E_0^2 C(x_0, y_0) e^{j\omega_0 \tau} \int_{-\infty}^{\infty} \int_{-\infty}^{\infty} \phi_d(x_4, y_4) \phi_a(x_4 - v\tau, y_4) dx_4 \, dy_4 \tag{7.20}$$

In this expression,

$$\phi_a(x, y) = \phi_\psi(x, y) \phi_g(x, y) \tag{7.21}$$

in which

$$\phi_g(\tau_x, \tau_y) = \int_{-\infty}^{\infty} \int_{-\infty}^{\infty} g^*(x, y) g(x + \tau_x, y + \tau_y) dx \, dy \tag{7.22}$$

is the two-dimensional autocorrelation function of the complex antenna pattern $g(x, y)$ given by Eq. (7.11).

We now substitute Eq. (7.20) in the Fourier transform expression, Eq. (6.14), to obtain an expression for the power density spectrum of the complex echo

$$\Phi_{cc}(\omega) = \frac{E_0^2 C(x_0, y_0)}{2\pi} \int_{-\infty}^{\infty} e^{-j(\omega - \omega_0)\tau} d\tau \int_{-\infty}^{\infty} \int_{-\infty}^{\infty} \phi_d(x_4, y_4)$$

$$\times \phi_a(x_4 - v\tau, y_4) \, dx_4 \, dy_4 \tag{7.23}$$

To put this general expression in a more manageable form, we first multiply and divide it by $e^{-j[(\omega - \omega_0)/v]x_4}$ and then let $x_5 = x_4 - v\tau$ to obtain

$$\Phi_{cc}(\omega) = \frac{E_0^2 C(x_0, y_0)}{2\pi v} \int_{-\infty}^{\infty} dy_4 \int_{-\infty}^{\infty} \phi_d(x_4, y_4) e^{-j[(\omega - \omega_0)/v]x_4} dx_4$$

$$\times \int_{-\infty}^{\infty} \phi_a(x_5, y_4) e^{-j[(\omega - \omega_0)/v]x_5} dx_5 \tag{7.24}$$

Now define the two-dimensional Fourier transform

$$\Phi_d(\omega_1, \omega_2) = \frac{1}{(2\pi)^2} \int_{-\infty}^{\infty} \int_{-\infty}^{\infty} \phi_d(x, y) e^{-j(\omega_1 x + \omega_2 y)} dx \, dy \tag{7.25}$$

Then,

$$\int_{-\infty}^{\infty} \Phi_d(\omega_1, \ \omega_2)e^{j\omega_2 y}\,d\omega_2 = \frac{1}{2\pi} \int_{-\infty}^{\infty} \phi_d(x, \ y)e^{-j\omega_1 x}\,dx \qquad (7.26)$$

We now substitute Eq. (7.26) in Eq. (7.24) to obtain

$$\Phi_{cc}(\omega) = \frac{E_0^2 C(x_0, \ y_0)}{v} \int_{-\infty}^{\infty} \Phi_d\left(\frac{\omega - \omega_0}{v}, \ \omega_2\right)d\omega_2 \cdots$$
$$\int_{-\infty}^{\infty}\int_{-\infty}^{\infty} \phi_a(x_5, \ y_4)e^{j[(\omega-\omega_0)/v]x_5}\,e^{j\omega_2 y_4}\,dx_5\,dy_4 \qquad (7.27)$$

This can be expressed in the form

$$\Phi_{cc}(\omega) = \frac{(2\pi)^2 E_0^2 C(x_0, \ y_0)}{v} \int_{-\infty}^{\infty} \Phi_d\left(\frac{\omega - \omega_0}{v}, \ \omega_2\right)\Phi_a\left(\frac{\omega_0 - \omega}{v}, \ -\omega_2\right)d\omega_2$$
$$(7.28)$$

in which

$$\Phi_a(\omega_1, \ \omega_2) = \frac{1}{(2\pi)^2} \int_{-\infty}^{\infty}\int_{-\infty}^{\infty} \phi_a(x, \ y)e^{-j(\omega_1 x + \omega_2 y)}\,dx\,dy \qquad (7.29)$$

Finally, for convenience, we define $\omega' = \omega - \omega_0$ to express Eq. (7.28) as

$$\Phi_{cc}(\omega_0 + \omega') = \frac{(2\pi)^2 E_0^2 C(x_0, \ y_0)}{v} \int_{-\infty}^{\infty} \Phi_d\left(\frac{\omega'}{v}, \ \omega_2\right)\Phi_a\left(-\frac{\omega'}{v}, \ -\omega_2\right)d\omega_2$$
$$(7.30)$$

This expression for the echo power density spectrum will now be interpreted in a manner similar to our discussion in Chapter 6.

V. Doppler Spectrum for Smooth and Periodic Terrain

The functions $\Phi_d(\omega_1, \ \omega_2)$ and $\Phi_a(\omega_1, \ \omega_2)$ are the two-dimensional versions of the functions defined by Eq. (6.33) and Eq. (6.34), respectively, and have similar interpretations. For example, smooth terrain is terrain in which the reflection coefficient does not vary with position so that $\phi_d(x, \ y) = |\gamma_0|^2$. Consequently, for smooth terrain,

$$\Phi_d(\omega_1, \ \omega_2) = |\gamma_0|^2 \delta(\omega_1)\delta(\omega_2) \qquad (7.31)$$

Substituting this result in Eq. (7.30), we obtain

$$\Phi_{cc}(\omega_0 + \omega') = (2\pi)^2 E_0^2 C(x_0, \ y_0)|\gamma_0|^2 \Phi_a(0, \ 0)\delta(\omega') \qquad (7.32)$$

The echo spectrum thus is an impulse at $\omega = \omega_0$ so that there is no Doppler effect. This phenominum is sometimes referred to as the smooth earth paradox. This result makes physical sense because the reflection from the terrain does not vary from place to place so that there is no information by which the radar can determine the velocity. In the general case of periodic terrain, the reflection coefficient is periodic with a fundamental period of T_x in the x direction and T_y in the y direction so that

$$\Phi_d(\omega_1, \omega_2) = \sum_{m=-\infty}^{\infty} \sum_{n=-\infty}^{\infty} A_{mn} \delta(\omega_1 + m\omega_x)\delta(\omega_2 + n\omega_y) \qquad (7.33)$$

in which $\omega_x = (2\pi/T_x)$ and $\omega_y = (2\pi/T_y)$. This could be an approximate model for a vehicle flying over water with a periodic surface structure. Substituting this expression in Eq. (7.30), we obtain

$$\Phi_{cc}(\omega_0 + \omega') = \frac{(2\pi)^2 E_0^2 C(x_0, y_0)}{v} \sum_{m=-\infty}^{\infty} \sum_{n=-\infty}^{\infty} A_{mn}$$

$$\times \int_{-\infty}^{\infty} \delta\left(\frac{\omega'}{v} + m\omega_x\right)\delta(\omega_2 + n\omega_y)\Phi_a\left(-\frac{\omega'}{v}, -\omega_2\right)d\omega_2 \quad (7.34)$$

$$= \frac{(2\pi)^2 E_0^2 C(x_0, y_0)}{v} \sum_{m=-\infty}^{\infty} \sum_{n=-\infty}^{\infty} A_{mn} \delta\left(\frac{\omega'}{v} + m\omega_x\right)\Phi_a\left(\frac{\omega'}{v}, n\omega_y\right)$$

$$= (2\pi)^2 E_0^2 C(x_0, y_0) \sum_{m=-\infty}^{\infty} \sum_{n=-\infty}^{\infty} A_{mn} \delta(\omega' + mv\omega_x)\Phi_a(-m\omega_x, n\omega_y)$$

$$(7.35)$$

Observe from this expression that the echo spectrum is nonzero only at the frequencies $\omega = \omega_0 + mv\omega_x$, $m = 0, \pm1, \pm2, \ldots$ so that the echo is a sinusoid with the same frequency as that of the transmitter modulated by a periodic wave with a fundamental angular frequency $\omega_m = v\omega_x$. Thus there is no Doppler effect for the case of periodic terrain. As in the case for smooth terrain, this result makes physical sense because the reflection from a periodic terrain is a periodic function so that the velocity v can be determined only if the fundamental period of the terrain in the direction of the velocity T_x is known.

VI.   Doppler Spectrum for Rough Terrain: General Expressions

We now consider rough terrain that, we shall see, is the type of terrain from which the Doppler effect is obtained. For rough terrain, $\phi_d(x, y)$ is a narrow pulse so that its Fourier transform $\Phi_d(\omega_1, \omega_2)$ is broad compared with $\Phi_a(\omega_1, \omega_2)$. Eq. (7.30) then is

$$\Phi_{cc}(\omega_0 + \omega') = \frac{(2\pi)^2 E_0^2 C(x_0, y_0)}{v} \Phi_d(0, 0) \int_{-\infty}^{\infty} \Phi_a\left(-\frac{\omega'}{v}, -\omega_2\right)d\omega_2 \quad (7.36)$$

In practice, the width of $\Phi_{cc}(\omega)$ is narrow as compared with the transmitter frequency ω_0. Thus, from Eq. (6.13), the echo spectrum about ω_0 is

$$\Phi_{rr}(\omega_0 + \omega') = \frac{(2\pi)^2 E_0^2 C(x_0,\, y_0)}{2v} \Phi_d(0,\, 0) \int_{-\infty}^{\infty} \Phi_a\left(-\frac{\omega'}{v},\, -\omega_2\right) d\omega_2 \quad (7.37)$$

To interpret this equation, we have from Eq. (7.21) that the two-dimensional autocorrelation function $\phi_a(x, y)$ is the product of $\phi_\psi(x, y)$ and $\phi_g(x, y)$. Parallel to our previous discussion in section 6.6, we'll discuss two cases: one in which $\phi_\psi(x, y)$ is a narrow pulse compared with $\phi_g(x, y)$ and the other in which $\phi_\psi(x, y)$ is a broad pulse compared with $\phi_g(x, y)$.

We first consider the case in which $\phi_\psi(x, y)$ is a narrow pulse compared with $\phi_g(x, y)$. Physically, $\phi_\psi(x, y)$ being a narrow pulse corresponds to the case in which the backscattering coefficient is a sensitive function of angle within the antenna beam so that the reflections at two slightly different angles within the antenna beam are essentially uncorrelated. Such a condition could occur at large values of the transmitter frequency ω_0 so that the wavelength is small compared with the irregularities of the individual scatterers.

If $\phi_\psi(x, y)$ is a narrow pulse compared with $\phi_g(x, y)$, we then observe from Eq. (7.21) that $\phi_a(x, y)$ also is a narrow pulse so that its Fourier transform $\Phi_a(\omega_1, \omega_2)$ is broad. For this condition, we then observe from Eq. (7.37) that the echo power density spectrum $\Phi_{rr}(\omega)$ will be broad, making it difficult to determine the presence of the Doppler effect. We thus can expect Doppler radars using optical lasers to encounter this difficulty over certain terrains.

We now consider the case in which $\phi_\psi(x, y)$ is a broad pulse compared with $\phi_g(x, y)$. Physically, this corresponds to the case in which the backscattering coefficient is not a sensitive function of angle within the antenna beam. Such a condition occurs for cases in which the wavelength is large compared with the irregularities of the individual scatterers. For this condition, we have from Eq. (7.21)

$$\phi_a(x,\, y) \approx \phi_\psi(0,\, 0)\phi_g(x,\, y) \quad (7.38)$$

so that its Fourier transform is

$$\Phi_a(\omega_1,\, \omega_2) \approx \phi_\psi(0,\, 0)\Phi_g(\omega_1,\, \omega_2) \quad (7.39)$$

From the two-dimensional Fourier transform, Eq. (7.22), this equation can be expressed as

$$\Phi_a(\omega_1,\, \omega_2) = (2\pi)^2 \phi_\psi(0,\, 0)|G(\omega_1,\, \omega_2)|^2 \quad (7.40)$$

in which

$$G(\omega_1,\, \omega_2) = \frac{1}{(2\pi)^2} \int_{-\infty}^{\infty} \int_{-\infty}^{\infty} g(x,\, y)e^{-j(\omega_1 x + \omega_2 y)} \, dx \, dy \quad (7.41)$$

Thus, for rough terrain in which Eq. (7.38) is valid, the general expression for the echo power density spectrum from Eq. (7.37) is

$$\Phi_{rr}(\omega_0 + \omega') = \frac{(2\pi)^4 E_0^2 C(x_0, y_0)}{2v} \Phi_d(0, 0)\Phi_\psi(0, 0)$$

$$\times \int_{-\infty}^{\infty} \left| G\left(-\frac{\omega'}{v}, -\omega_2\right) \right|^2 d\omega_2 \tag{7.42}$$

This equation is the two-dimensional expression parallel to Eq. (6.47).

VII.   Analysis of Some Special Cases

To gain some understanding of Eq. (7.42), we shall interpret this expression by considering some special cases. For analytical simplicity, we'll ignore the inverse-square law and analyze the echo spectrum for the case in which $b(x, y)$, defined by Eq. (7.9), is a constant b_0 over the region for which the antenna pattern $s(x, y)$ is significantly different from zero. For this case, we have from Eq. (7.11)

$$g(x, y) = b_0\, p(x, y) \tag{7.43}$$

in which $p(x, y)$ is the complex antenna pattern given by Eq. (7.10). From Eq. (7.41), the Fourier transform of Eq. (7.43) is

$$G(-\omega_1, -\omega_2) = b_0\, P(\omega_1, \omega_2) \tag{7.44}$$

in which

$$P(\omega_1, \omega_2) = \frac{1}{(2\pi)^2} \int_{-\infty}^{\infty} \int_{-\infty}^{\infty} p(x, y)e^{j(\omega_1 x + \omega_2 y)}\, dx\, dy \tag{7.45}$$

is the Fourier transform of the complex antenna pattern. We then have from Eq. (7.42)

$$\Phi_{rr}(\omega_0 + \omega') = \frac{(2\pi)^2 b_0^2 E_0^2 C(x_0, y_0)}{2v} \Phi_d(0, 0)\Phi_\psi(0, 0)$$

$$\times \int_{-\infty}^{\infty} \left| P\left(\frac{\omega'}{v}, \omega_2\right) \right|^2 d\omega_2 \tag{7.46}$$

We now interpret this expression for some specific complex antenna patterns $p(x, y)$.

A.   The Doppler Spectrum for a Narrow Antenna Pattern

From Eq. (7.10) we have

$$P\left(\frac{\omega'}{v}, \omega_2\right) = \frac{1}{(2\pi)^2} \int_{-\infty}^{\infty} \int_{-\infty}^{\infty} s(x, y)e^{-j[(4\pi/\lambda)r - (\omega/v)x - \omega_2 y)]}\, dx\, dy \tag{7.47}$$

in which $r^2 = h^2 + x^2 + y^2$. We first consider the case of a thin and narrow antenna pattern centered at $(x, y) = (x_0, y_0)$ as shown in Fig. 7.2. By a thin and narrow antenna pattern, I mean $s(x, y) \approx 0$ for $|x - x_0| > \delta_x$ and $|y - y_0| > \delta_y$ in which $\delta_x^2 + \delta_y^2 \ll 2(x_0 \delta_x + y_0 \delta_y)$ so that, in the range for which $s(x, y)$ is significantly different from zero, r can be approximated as

$$r = [h^2 + x^2 + y^2]^{1/2}$$

$$= [h^2 + x_0^2 + y_0^2 + 2x_0(x - x_0) + 2(y - y_0) + (x - x_0)^2 + (y - y_0)^2]^{1/2}$$

$$\approx [r_0^2 + 2x_0(x - x_0) + 2y_0(y - y_0)]^{1/2}$$

$$\approx r_0 + \frac{x_0(x - x_0)}{r_0} + \frac{y_0(y - y_0)}{r_0} = \frac{h^2}{r_0} + \frac{x x_0}{r_0} + \frac{y y_0}{r_0} \tag{7.48}$$

We then can approximate Eq. (7.47) as

$$P\left(\frac{\omega'}{v}, \omega_2\right) = e^{-j(4\pi/\lambda)(h^2/r_0)} \frac{1}{(2\pi)^2} \int_{-\infty}^{\infty} \int_{-\infty}^{\infty} s(x, y)$$

$$\times e^{j((\omega'/v) - (4\pi/\lambda)(x_0/r_0))x} e^{j(\omega_2 - (4\pi/\lambda)(y_0/r_0))y} \, dx \, dy$$

$$= e^{-j(4\pi/\lambda)(h^2/r_0)} S\left[\frac{\omega'}{v} - \frac{4\pi x_0}{\lambda r_0}, \omega_2 - \frac{4\pi y_0}{\lambda r_0}\right] \tag{7.49}$$

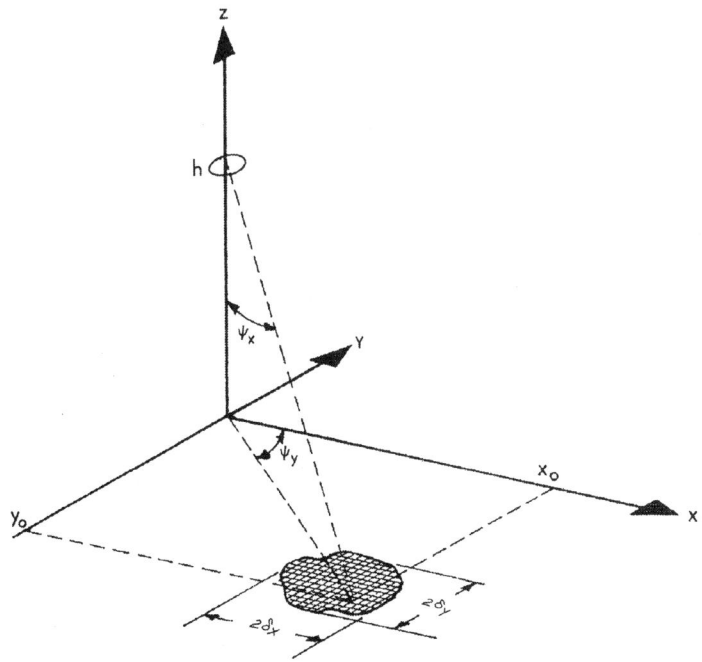

Fig. 7.2   Geometric representation for a thin and narrow antenna pattern.

in which

$$S[\omega_1, \ \omega_2] = \frac{1}{(2\pi)^2} \int_{-\infty}^{\infty} \int_{-\infty}^{\infty} s(x, \ y) e^{j(\omega_1 x + \omega_2 y)} \, dx dy \qquad (7.50)$$

is the Fourier transform of the two-way antenna pattern as projected on the terrain.

For tractability, we consider the case in which $s(x, y)$ is separable, which means it can be expressed as the product

$$s(x, \ y) = s_x(x) s_y(y) \qquad (7.51)$$

so that from Eq. (7.50)

$$S[\omega_1, \ \omega_2] = S_x(\omega_1) S_y(\omega_2) \qquad (7.52)$$

in which, for $z = x$ or y,

$$S_z(\omega) = \frac{1}{2\pi} \int_{-\infty}^{\infty} s_z(z) e^{j\omega z} \, dz \qquad (7.53)$$

Then, from Eq. (7.49),

$$P\left(\frac{\omega'}{v}, \ \omega_2\right) = e^{-j(4\pi/\lambda)(h^2/r_0)} S_x\left(\frac{\omega'}{v} - \frac{4\pi x_0}{\lambda \ r_0}\right) S_y\left(\omega_2 - \frac{4\pi y_0}{\lambda \ r_0}\right) \qquad (7.54)$$

Substituting into Eq. (7.46) we obtain

$$\Phi_{rr}(\omega_0 + \omega') = \frac{(2\pi)^2 b_0^2 E_0^2 C(x_0, \ y_0)}{2v} \Phi_d(0, \ 0) \Phi_\psi(0, \ 0) C_y^2 \left| S_x\left(\frac{\omega'}{v} - \frac{4\pi x_0}{\lambda \ r_0}\right) \right|^2 \qquad (7.55)$$

in which

$$C_y^2 = \int_{-\infty}^{\infty} |S_y(\omega)|^2 \, d\omega \qquad (7.56)$$

By Parseval's theorem, this constant also is equal to

$$C_y^2 = \frac{1}{2\pi} \int_{-\infty}^{\infty} |s_y(y)|^2 \, dy \qquad (7.57)$$

Observe from Eq. (7.55) that the shape of the power density spectrum of the echo is the square of the Fourier transform of the x component (the component in the direction of travel) of the two-way antenna pattern. However, the transform is centered about the frequency $\omega = \omega_0 + \omega_d$ in which, from

Eq. (7.55') and Fig. 7.2,

$$\omega_d = \frac{4\pi v\, x_0}{\lambda} \frac{x_0}{r_0} = \frac{4\pi v}{\lambda} \sin \psi_x \cos \psi_y \text{ rad/s}$$

or

$$f_d = \frac{1}{2\pi}\,\omega_d = \frac{2v}{\lambda} \sin \psi_x \cos \psi_y \text{ hertz} \qquad (7.58)$$

B.   The Doppler Spectrum for a Gaussian Antenna Pattern

We now consider the case of a Gaussian antenna pattern for which

$$s(x,\ y) = \frac{1}{2\pi\alpha_x\alpha_y}\,e^{-((x-x_0)^2/2a_x^2)}e^{-((y-y_0)^2/2a_y^2)} \qquad (7.59)$$

To perform the required integration, we approximate r in Eq. (7.47) similar to that used in Appendix A:

$$r = [h^2 + x^2 + y^2]^{1/2}$$
$$= [h^2 + x_0^2 + y_0^2 + (x^2 - x_0^2) + (y^2 - y_0^2)]^{1/2}$$
$$= [r_0^2 + (x^2 - x_0^2) + (y^2 - y_0^2)]^{1/2}$$
$$\approx r_0\left[1 + \frac{(x^2 - x_0^2)}{2r_0^2} + \frac{(y^2 - y_0^2)}{2r_0^2}\right] \qquad (7.60)$$

in which $r_0 = [h^2 + x_0^2 + y_0^2]^{1/2}$ is the radial distance from the antenna to the beam center on the terrain. As in Appendix A, the approximation error is less than one percent for

$$\frac{(x^2 - x_0^2)}{r_0^2} + \frac{(y^2 - y_0^2)}{r_0^2} < 0.326 \qquad (7.61)$$

Now substitute the approximation Eq. (7.60) in Eq. (7.47) to obtain

$$P\left(\frac{\omega'}{v},\ \omega_2\right) = \frac{1}{2\pi}\frac{1}{a_x\sqrt{2\pi}}e^{-j(4\pi/\lambda)r_0}$$

$$\times \int_{-\infty}^{\infty} e^{-((x-x_0)^2/2a_x^2)}e^{-j(4\pi/\lambda)(x^2-x_0^2/2r_0)}e^{j(\omega'/v)x}\,dx$$

$$\times \frac{1}{a_y\sqrt{2\pi}}\int_{-\infty}^{\infty} e^{-((y-y_0)^2/2a_y^2)}e^{-j(4\pi/\lambda)(y^2-y_0^2/2r_0)}e^{j\omega_2 y}\,dy \qquad (7.62)$$

We now use the results obtained in Appendix A to express this equation as

$$P\left(\frac{\omega'}{v}, \omega_2\right) = \frac{1}{2\pi}\frac{1}{a_x\sqrt{2\pi}}\frac{1}{a_y\sqrt{2\pi}}e^{-j(4\pi/\lambda)r_0}e^{j(\omega'/v)x_0}e^{j\omega_2 y_0}$$

$$\times F_x\left[\frac{\omega'}{v} - \frac{4\pi x_0}{\lambda}\frac{}{r_0}\right]F_y\left[\omega_2 - \frac{4\pi y_0}{\lambda}\frac{}{r_0}\right] \tag{7.63}$$

in which $F(\omega)$ is given by Eq. (A.15). We then obtain the echo power density spectrum by substituting Eq. (7.63) in Eq. (7.46)

$$\Phi_{rr}(\omega_0 + \omega') = \frac{b_0^2 E_0^2 C(x_0, y_0)}{2a_x a_y v}\Phi_d(0, 0)\Phi_\psi(0, 0)C_y^2\left|F_x\left(\frac{\omega'}{v} - \frac{4\pi x_0}{\lambda}\frac{}{r_0}\right)\right|^2 \tag{7.64a}$$

in which

$$C_y^2 = \int_{-\infty}^{\infty} |F_y(\omega)|^2 d\omega = \frac{2\pi^{3/2}a_y}{\sigma_y} \tag{7.64b}$$

and

$$\sigma_y = \frac{1}{(2a_y^2)^2} + \left(\frac{2\pi}{\lambda r_0}\right)^2 \tag{7.64c}$$

By rationalizing $F_x(\omega)$ as was done for Eq. (A.19), this expression can be expressed in the simplified form

$$\Phi_{rr}(\omega_0 + \omega') = Ke^{-((\omega-\omega_d)^2/2\sigma_x^2)} \tag{7.65}$$

in which K is a constant, ω_d is given by Eq. (7.58), and

$$2\sigma_x^2 = \left(\frac{v}{a_x}\right)^2\left[1 + \left(\frac{4\pi a_x^2}{\lambda r_0}\right)^2\right] \tag{7.66}$$

which is similar to the bandwidth expression given by Eq. (6.71b). We observe that for this case, too, a change in the width of the antenna pattern in the direction perpendicular to the direction of travel, the y direction, changes the amplitude of the spectrum but not its shape or the Doppler frequency. The reason is that with the approximation given by Eq. (7.60), $P(\omega_1, \omega_2)$ is separable. In fact, we observe that for any case in which $P(\omega_1, \omega_2)$ can be expressed in separable form as $P(\omega_1, \omega_2) = P_x(\omega_1) P_y(\omega_2)$, the resulting echo power density spectrum will have the form $\Phi_{rr}(\omega + \omega') = KP_x(\omega_1)$. For such cases, the shape of the echo spectrum depends only on the shape of the antenna pattern in the direction of travel, the x direction. The shape of the antenna pattern in the direction perpendicular to the direction of travel, the y direction, affects only the echo spectrum amplitude.

Chapter 8

Center Frequency and Bandwidth of the Doppler Spectrum

IN THE last chapter, the general expressions for the power density spectrum of the echo from a continuous wave (CW) airborne Doppler radar were obtained. Those expressions allow the determination of the exact shape of the spectrum from knowledge of certain statistical properties of the terrain and the antenna pattern as projected on the terrain. Although approximate analytical expressions were obtained for certain special cases, a computer would generally have to be used for their evaluation. Fortunately, the exact shape of the spectrum is not always required. Often only the Doppler frequency and the Doppler spectrum bandwidth are required. We have seen from our approximate analysis that the center frequency of the spectrum is what we call the Doppler frequency. It can be tracked experimentally by comparing the outputs of a low-pass and a high-pass filter. Thus for the determination of the ground velocity, only the center frequency of the Doppler spectrum and its relation to the vehicle ground velocity are required.

However, the measurement of the Doppler spectrum center frequency will be in error; the larger the spectrum bandwidth, the larger is this measurement error. Thus the spectrum bandwidth is another parameter that is desired. With these data, we can determine the ratio of the spectrum center frequency to the spectrum bandwidth, which I call the Doppler spectrum quality factor Q. The Doppler quality factor Q thus is a measure of the accuracy of the measurement of the spectrum center frequency. An objective of Doppler radar design should be the design of a system with as large a quality factor Q as reasonable. It would be inefficient to first calculate the exact spectrum from which these quantities can be determined. Rather, specific equations are needed from which these quantities can be determined without the necessity of first determining the exact Doppler spectrum. In this chapter, the required specific expressions are derived using the theory developed in the last chapter. These expressions first will be derived in terms of the antenna pattern as projected on the terrain. However, because an antenna pattern is normally expressed in coordinates relative to the antenna, the expressions then will be transformed into ones that are in terms of the antenna coordinates.

I.   Center Frequency and Bandwidth of a Spectrum

To begin, the center frequency ω_c of the Doppler power density spectrum is the frequency for which one-half of the average power is below ω_c. Mathematically, this is the first moment of the power density spectrum, which is

$$\omega_c = \frac{\int_0^\infty \omega \Phi_{rr}(\omega)d\omega}{\int_0^\infty \Phi_{rr}(\omega)d\omega} \tag{8.1}$$

A quantity of more direct interest is the displacement of the Doppler spectrum center frequency from the transmitter frequency ω_0. We thus define the Doppler frequency as

$$\delta = \omega_c - \omega_0 \tag{8.2}$$

We then can express Eq. (8.1) as

$$\delta = \frac{\int_{-\omega_0}^\infty \omega' \Phi_{rr}(\omega_0 + \omega')d\omega'}{\int_{-\omega_0}^\infty \Phi_{rr}(\omega_0 + \omega')d\omega'} \tag{8.3}$$

in which $\omega' = \omega - \omega_0$.

The second moment of a positive curve is a measure of its width. Because the power density spectrum is a positive curve, we define the spectrum bandwidth as σ in which

$$\sigma^2 = \frac{\int_0^\infty (\omega - \omega_c)^2 \Phi_{rr}(\omega)d\omega}{\int_0^\infty \Phi_{rr}(\omega)d\omega} \tag{8.4}$$

This is the second moment of the power density spectrum about its center frequency.[1] With the change of variable $\omega' = \omega - \omega_0$ we then have

$$\sigma^2 = \frac{\int_{-\omega_0}^\infty (\omega' - \delta)^2 \Phi_{rr}(\omega_0 + \omega')d\omega'}{\int_{-\omega_0}^\infty \Phi_{rr}(\omega_0 + \omega')d\omega'} \tag{8.5}$$

By substituting $(\omega' - \delta)^2 = (\omega')^2 - 2\delta\omega' + \delta^2$ and using Eq. (8.3), we can express Eq. (8.5) as

$$\sigma^2 + \delta^2 = \frac{\int_{-\omega_0}^\infty (\omega')^2 \Phi_{rr}(\omega_0 + \omega')d\omega'}{\int_{-\omega_0}^\infty \Phi_{rr}(\omega_0 + \omega')d\omega'} \tag{8.6}$$

---

[1] There are many definitions of bandwidth. This definition of bandwidth is a measure of the dispersion of the Doppler spectrum about the Doppler frequency.

Equation (8.3) and Eq. (8.6) are the general expressions we'll use to determine the Doppler frequency δ and Doppler spectrum bandwidth σ.

II.   Expressions for the Doppler Spectrum using Terrain Coordinates

The Doppler power density spectrum is required to evaluate Eq. (8.3) and Eq. (8.6). To eliminate this need, we shall transform these equations by using the results obtained in Chapter 7 for rough terrain. We'll first collect the needed previous results.

The general expression for the single-sided Doppler power density spectrum for rough terrain is, from Eq. (7.37)

$$\Phi_{rr}(\omega_0 + \omega') = \frac{(2\pi)^2 E_0^2 C(x_0, y_0)}{2v} \Phi_d(0, 0) \int_{-\infty}^{\infty} \Phi_a\left(-\frac{\omega'}{v}, -\omega_2\right) d\omega_2 \quad (8.7a)$$

Also, for the case in which the random component of the terrain backscattering coefficient is not a sensitive function of angle over the range of angles within the antenna beam so that $\phi_\psi(x, y)$ is a broad pulse compared with $\phi_g(x, y)$, we have from Eq. (7.40) that

$$\Phi_a(\omega_1, \omega_2) = (2\pi)^2 \phi_\psi(0, 0)|G(\omega_1, \omega_2)|^2 \quad (8.7b)$$

in which, from Eq. (7.41)

$$G(\omega_1, \omega_2) = \frac{1}{(2\pi)^2} \int_{-\infty}^{\infty} \int_{-\infty}^{\infty} g(x, y) e^{-j(\omega_1 x + \omega_2 y)} \, dx \, dy \quad (8.8)$$

and $g(x, y)$ is the effect complex two-way antenna pattern. From Eq. (7.10) and Eq. (7.11) it can be expressed as

$$g(x, y) = k(x, y) e^{-j(4\pi r/\lambda)} \quad (8.9)$$

in which the wavelength is

$$\lambda = \frac{2\pi c}{\omega_0} \quad (8.10a)$$

and the distance from the antenna to a point on the ground is

$$r^2 = h^2 + x^2 + y^2 \quad (8.10b)$$

with ω_0 being the radiated sinusoidal frequency and c the velocity of light, $c \approx 3 \cdot 10^8$ m/s. Also

$$k(x, y) = b(x, y) s(x, y) \quad (8.11)$$

in which $s(x, y)$ is the two-way antenna pattern as projected on the terrain and $b(x, y)$ is the expected value of the terrain reflection coefficient; for diffuse surfaces, this would correspond to the Lambert law.

We begin our transformation by substituting Eq. (8.7a) in Eq. (8.3) and Eq. (8.6) to obtain

$$\delta = \frac{\int_{-\omega_0}^{\infty} \omega' d\omega' \int_{-\infty}^{\infty} \Phi_a((-\omega'/v), -\omega_2) d\omega_2}{\int_{-\omega_0}^{\infty} d\omega' \int_{-\infty}^{\infty} \Phi_a((-\omega'/v), -\omega_2) d\omega_2} \tag{8.12}$$

and

$$\sigma^2 + \delta^2 = \frac{\int_{-\omega_0}^{\infty} (\omega')^2 d\omega' \int_{-\infty}^{\infty} \Phi_a((-\omega'/v), -\omega_2) d\omega_2}{\int_{-\omega_0}^{\infty} d\omega' \int_{-\infty}^{\infty} \Phi_a((-\omega'/v), -\omega_2) d\omega_2} \tag{8.13}$$

Note from Eq. (6.13) that we are using the general expression for the single-sided power density spectrum. Furthermore, because the spectrum bandwidth is narrow compared with the radiated frequency ω_0, the lower limit in the previous two expressions can be extended to negative infinity without altering the value of the integrals. Then, with the substitution of Eq. (8.7b) and the change of variable $\omega_1 = \omega'/v$, the last two equations can be expressed as

$$\delta = \frac{v \int_{-\infty}^{\infty} \omega_1 d\omega_1 \int_{-\infty}^{\infty} |G(-\omega_1, -\omega_2)|^2 d\omega_2}{\int_{-\infty}^{\infty} d\omega_1 \int_{-\infty}^{\infty} |G(-\omega_1, -\omega_2)|^2 d\omega_2} \tag{8.14}$$

and

$$\sigma^2 + \delta^2 = \frac{v^2 \int_{-\infty}^{\infty} \omega_1^2 d\omega_1 \int_{-\infty}^{\infty} |G(-\omega_1, -\omega_2)|^2 d\omega_2}{\int_{-\infty}^{\infty} d\omega_1 \int_{-\infty}^{\infty} |G(-\omega_1, -\omega_2)|^2 d\omega_2} \tag{8.15}$$

We now use the Parseval relations derived in Appendix C to express these two equations in terms of $g(x, y)$. By substituting Eq. (C.6), Eq. (C.9), and Eq. (C.10) in Eq. (8.14) and Eq. (8.15) we obtain

$$\delta = \frac{jv \int_{-\infty}^{\infty} \int_{-\infty}^{\infty} g^*(x, y) \frac{\partial}{\partial x} g(x, y) dx dy}{\int_{-\infty}^{\infty} \int_{-\infty}^{\infty} |g(x, y)|^2 dx dy} \tag{8.16}$$

and

$$\sigma^2 + \delta^2 = \frac{v^2 \int_{-\infty}^{\infty} \int_{-\infty}^{\infty} \left| \frac{\partial}{\partial x} g(x, y) \right|^2 dx dy}{\int_{-\infty}^{\infty} \int_{-\infty}^{\infty} |g(x, y)|^2 dx dy} \tag{8.17}$$

To simplify these expressions, we first note from Eq. (8.9) that with the use of Eqs. (8.10),

$$\frac{\partial}{\partial x}g(x,y) = \left[\frac{\partial}{\partial x}k(x,y) - j\frac{4\pi x}{\lambda r}k(x,y)\right]e^{-j(4\pi r/\lambda)} \tag{8.18}$$

Making use of the fact that $k(x,y)$ is a real function, we then have from Eq. (8.9) and Eq. (8.18)

$$|g(x,y)|^2 = k^2(x,y) \tag{8.19}$$

$$\left|\frac{\partial}{\partial x}g(x,y)\right|^2 = \left[\frac{\partial}{\partial x}k(x,y)\right]^2 + \left[\frac{4\pi x}{\lambda r}k(x,y)\right]^2 \tag{8.20}$$

and

$$g^*(x,y)\frac{\partial}{\partial x}g(x,y) = k(x,y)\frac{\partial}{\partial x}k(x,y) - j\frac{4\pi x}{\lambda r}k^2(x,y) \tag{8.21}$$

Before substituting these expressions in Eq. (8.16) and Eq. (8.17), observe in Eq. (8.21) that $k(x,y)(\partial/\partial x)k(x,y) = 1/2\ (\partial/\partial x)\ k^2(x,y)$. Thus

$$\int_{-\infty}^{\infty}\int_{-\infty}^{\infty}k(x,y)\frac{\partial}{\partial x}k(x,y)dxdy = \frac{1}{2}\int_{-\infty}^{\infty}\int_{-\infty}^{\infty}\frac{\partial}{\partial x}k^2(x,y)dxdy$$

$$= \frac{1}{2}\int_{-\infty}^{\infty}\left[k^2(\infty,y) - k^2(-\infty,y)\right]dy$$

$$= 0 \tag{8.22}$$

The value of the integral is zero because there is no reflection at the horizon so that, from Eq. (8.11), $k(\infty,y)$ and $k(-\infty,y)$ are zero. Thus, by substituting Eq. (8.19), Eq. (8.20), and Eq. (8.21) in Eq. (8.16) and Eq. (8.17) we obtain, with the use of the result given by Eq. (8.22)

$$\delta = \frac{4\pi v}{\lambda}\frac{\int_{-\infty}^{\infty}\int_{-\infty}^{\infty}(x/r)k^2(x,y)dxdy}{\int_{-\infty}^{\infty}\int_{-\infty}^{\infty}k^2(x,y)dxdy} \tag{8.23}$$

and

$$\sigma^2 + \delta^2 = \frac{v^2\int_{-\infty}^{\infty}\int_{-\infty}^{\infty}[(\partial/\partial x)k(x,y)]^2dxdy + (4\pi v/\lambda)^2\int_{-\infty}^{\infty}\int_{-\infty}^{\infty}[(x/r)k(x,y)]^2dxdy}{\int_{-\infty}^{\infty}\int_{-\infty}^{\infty}k^2(x,y)dxdy} \tag{8.24}$$

The integrals in the last two equations have values that depend only on $k(x,y)$. From Eq. (8.11), this function is the amplitude of the effective two-way antenna pattern. We thus observe from Eq. (8.23) that δ, the displacement of the Doppler spectrum center frequency from the transmitted frequency, is proportional to v/λ irrespective of the antenna geometry or the expected terrain backscatter. We also observe from Eq. (8.24) that the spectrum bandwidth σ is proportional to the velocity v irrespective of the antenna geometry or the expected terrain backscatter. Thus the Doppler spectrum quality factor $Q = (\delta/2\sigma)$ is invariant with the ground velocity of the vehicle.

III. Transformation from Terrain to Antenna Coordinates

Equation (8.23) and Eq. (8.24) for δ and σ are in terms of $k(x, y)$, the effective two-way antenna pattern as projected on the terrain. However, the antenna pattern $s(x, y)$ and the expected terrain backscatter $b(x, y)$ are normally expressed in coordinates relative to the antenna. We thus need to express Eq. (8.23) and Eq. (8.24) in terms of the effective two-way antenna pattern expressed in antenna coordinates.

Figure 8.1 is a schematic for defining the geometry we use to obtain the desired coordinate transformation. As shown, the antenna is a distance h above

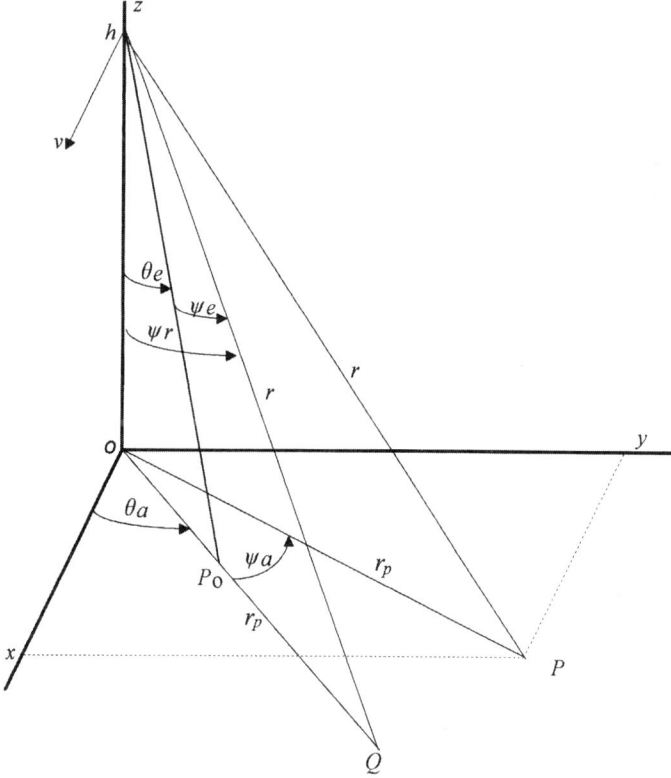

Fig. 8.1   Geometry used for transforming from rectangular to antenna co-ordinates.

the average terrain surface and is traveling in the x direction at a constant velocity v. Relative to the (x, y, z) coordinate system shown, the antenna is at the point $(0, 0, h)$. The antenna beam center on the terrain is at the point P_0. The line from $(0, 0, h)$ to the point P_0 thus is the antenna beam axis. The antenna is gimbaled so that it can be rotated in azimuth and elevation. The antenna shown in Fig. 8.1 has been rotated θ_a radians in azimuth and θ_e radians in elevation. Also, ψ_r is the range angle and ψ_a is the azimuth angle from the antenna to a point P on the terrain.

There are four integrals in Eq. (8.23) and Eq. (8.24). They are:

$$I_0 = \int_{-\infty}^{\infty} \int_{-\infty}^{\infty} k^2(x, y) dx dy \tag{8.25}$$

$$I_1 = \int_{-\infty}^{\infty} \int_{-\infty}^{\infty} \left(\frac{x}{r}\right) k^2(x, y) dx dy \tag{8.26}$$

$$I_2 = \int_{-\infty}^{\infty} \int_{-\infty}^{\infty} \left(\frac{x}{r}\right)^2 k^2(x, y) dx dy \tag{8.27}$$

$$I_3 = \int_{-\infty}^{\infty} \int_{-\infty}^{\infty} \left[\frac{\partial}{\partial x} k(x, y)\right]^2 dx dy \tag{8.28}$$

We thus must express each integral in antenna coordinates. We begin by expressing $k(x, y)$ in the antenna angular coordinates as

$$k(x, y) = \frac{1}{r^2} K(\psi_r, \psi_a) \tag{8.29}$$

The factor $1/r^2$ is a consequence of $k(x, y)$ being a two-way effective antenna pattern. Note that the effective two-way antenna pattern in angular coordinates is $K(\psi_r, \psi_a) = S(\psi_r, \psi_a) B(\psi_r, \psi_a)$ in which $S(\psi_r, \psi_a)$ is the two-way antenna pattern in angular coordinates and $B(\psi_r, \psi_a)$ is the expected value of the terrain backscatter in angular coordinates. For diffuse surfaces, for example, it would correspond to the Lambert law.

To express the integrals in the desired angular coordinates, we need to determine the Jacobian of the transformation and to express x and r in angular coordinates. For this, consider a point P as shown in Fig. 8.1. The point is on the $x-y$ plane at a distance r from the antenna. We then obtain the following relations from the right triangle h-o-Q with the hypotenuse of length r

$$r = h \sec \psi_r \tag{8.30a}$$

and

$$r_p = r \sin \psi_r \tag{8.30b}$$

Also from Fig. 8.1

$$x = r_p \cos (\theta_a + \psi_a) \tag{8.31a}$$

and

$$y = r_p \sin(\theta_a + \psi_a) \tag{8.31b}$$

Substituting Eqs. (8.30) in Eqs. (8.31), we obtain the following relations

$$r = h \sec \psi_r \tag{8.32a}$$

$$x = h \tan \psi_r \cos(\theta_a + \psi_a) \tag{8.32b}$$

and

$$y = h \tan \psi_r \sin(\theta_a + \psi_a) \tag{8.32c}$$

Now, the Jacobian of the transformation[2] is equal to the determinant

$$J = \frac{\partial(x, y)}{\partial(\psi_r, \psi_a)} = \begin{vmatrix} \dfrac{\partial x}{\partial \psi_r} & \dfrac{\partial x}{\partial \psi_a} \\ \dfrac{\partial y}{\partial \psi_r} & \dfrac{\partial y}{\partial \psi_a} \end{vmatrix}$$

$$= \frac{\partial x}{\partial \psi_r}\frac{\partial y}{\partial \psi_a} - \frac{\partial x}{\partial \psi_a}\frac{\partial y}{\partial \psi_r} \tag{8.33a}$$

With the use of Eqs. (8.32), its value is

$$J = h^2 \sec^2 \psi_r \tan \psi_r \cos^2(\theta_a + \psi_a) + h^2 \sec^2 \psi_r \tan \psi_r \sin^2(\theta_a + \psi_a)$$
$$= h^2 \sec^2 \psi_r \tan \psi_r \tag{8.33b}$$

The integrals I_0, I_1, and I_2 now can be expressed in antenna angular coordinates as

$$I_0 = \int_{-\pi/2}^{\pi/2} \int_{-\pi/2}^{\pi/2} \frac{J}{r^4} K^2(\psi_r, \psi_a) d\psi_r d\psi_a \tag{8.34a}$$

$$I_1 = \int_{-\pi/2}^{\pi/2} \int_{-\pi/2}^{\pi/2} \frac{xJ}{r^5} K^2(\psi_r, \psi_a) d\psi_r d\psi_a \tag{8.34b}$$

$$I_2 = \int_{-\pi/2}^{\pi/2} \int_{-\pi/2}^{\pi/2} \frac{x^2 J}{r^6} K^2(\psi_r, \psi_a) d\psi_r d\psi_a \tag{8.34c}$$

---

[2] The Jacobian is a factor that accounts for any changes of scale when transforming from one coordinate system to another.

in which r, x, and J are given in antenna angular coordinates by Eq. (8.32a), Eq. (8.32b), and Eq. (8.33b), respectively.

We now need to transform integral I_3. For this, we have from the differentiation chain rule that

$$\frac{\partial}{\partial x} k(x, y) = \frac{\partial}{\partial \psi_r} k(x, y) \frac{\partial \psi_r}{\partial x} + \frac{\partial}{\partial \psi_a} k(x, y) \frac{\partial \psi_a}{\partial x} \tag{8.35}$$

To evaluate this expression, we first note that Eq. (8.32b) and Eq. (8.32c) are in the form

$$\left. \begin{array}{l} x = f_x(\psi_r, \psi_a) \\ y = f_y(\psi_r, \psi_a) \end{array} \right\} \tag{8.36}$$

The derivative of these equations with respect to x is

$$\left. \begin{array}{l} 1 = \dfrac{\partial f_x}{\partial \psi_r} \dfrac{\partial \psi_r}{\partial x} + \dfrac{\partial f_x}{\partial \psi_a} \dfrac{\partial \psi_a}{\partial x} \\[2mm] 0 = \dfrac{\partial f_y}{\partial \psi_r} \dfrac{\partial \psi_r}{\partial x} + \dfrac{\partial f_y}{\partial \psi_a} \dfrac{\partial \psi_a}{\partial x} \end{array} \right\} \tag{8.37}$$

Thus, with the use of determinants, we have

$$\frac{\partial \psi_r}{\partial x} = \frac{\begin{vmatrix} 1 & \dfrac{\partial f_x}{\partial \psi_a} \\[2mm] 0 & \dfrac{\partial f_y}{\partial \psi_a} \end{vmatrix}}{\begin{vmatrix} \dfrac{\partial f_x}{\partial \psi_r} & \dfrac{\partial f_x}{\partial \psi_a} \\[2mm] \dfrac{\partial f_y}{\partial \psi_r} & \dfrac{\partial f_y}{\partial \psi_a} \end{vmatrix}} \tag{8.38}$$

The value of the determinant in the denominator is recognized as the value of the Jacobian J given by Eq. (8.33a) so that Eq. (8.38) can be expressed as:

$$\frac{\partial \psi_r}{\partial x} = \frac{1}{J} \frac{\partial f_y}{\partial \psi_a} = \frac{1}{J} \frac{\partial y}{\partial \psi_a} \tag{8.39a}$$

In a similar manner, we obtain from Eq. (8.37)

$$\frac{\partial \psi_a}{\partial x} = \frac{1}{J} \begin{vmatrix} \dfrac{\partial f_x}{\partial \psi_r} & 1 \\[2mm] \dfrac{\partial f_y}{\partial \psi_r} & 0 \end{vmatrix} = -\frac{1}{J} \frac{\partial f_y}{\partial \psi_r} = -\frac{1}{J} \frac{\partial y}{\partial \psi_r} \tag{8.39b}$$

By substituting Eq. (8.39a) and Eq. (8.39b) in Eq. (8.35), we then obtain

$$\frac{\partial k(x, y)}{\partial x} = \frac{1}{J} \left[\frac{\partial k(x, y)}{\partial \psi_r} \frac{\partial y}{\partial \psi_a} - \frac{\partial k(x, y)}{\partial \psi_a} \frac{\partial y}{\partial \psi_r} \right] \tag{8.40}$$

Now, from Eq. (8.29), we have

$$\left. \begin{array}{l} \dfrac{\partial k(x, y)}{\partial \psi_r} = \dfrac{1}{r^2} \dfrac{\partial K(\psi_r, \psi_a)}{\partial \psi_r} - \dfrac{2K(\psi_r, \psi_a)}{r^3} \dfrac{\partial r}{\partial \psi_r} \\[3mm] \dfrac{\partial k(x, y)}{\partial \psi_a} = \dfrac{1}{r^2} \dfrac{\partial K(\psi_r, \psi_a)}{\partial \psi_a} - \dfrac{2K(\psi_r, \psi_a)}{r^3} \dfrac{\partial r}{\partial \psi_a} \end{array} \right\} \tag{8.41}$$

We now substitute Eq. (8.41) in Eq. (8.40) to obtain

$$\begin{aligned} \frac{\partial k(x, y)}{\partial x} &= \frac{1}{J} \left[\frac{1}{r^2} \frac{\partial K(\psi_r, \psi_a)}{\partial \psi_r} - \frac{2K(\psi_r, \psi_a)}{r^3} \frac{\partial r}{\partial \psi_r} \right] \frac{\partial y}{\partial \psi_a} \\ &\quad - \frac{1}{J} \left[\frac{1}{r^2} \frac{\partial K(\psi_r, \psi_a)}{\partial \psi_a} - \frac{2K(\psi_r, \psi_a)}{r^3} \frac{\partial r}{\partial \psi_a} \right] \frac{\partial y}{\partial \psi_r} \\ &= \frac{1}{Jr^2} \left[\frac{\partial K(\psi_r, \psi_a)}{\partial \psi_r} \frac{\partial y}{\partial \psi_a} - \frac{\partial K(\psi_r, \psi_a)}{\partial \psi_a} \frac{\partial y}{\partial \psi_r} \right] \\ &\quad + \frac{2K(\psi_r, \psi_a)}{Jr^3} \left[\frac{\partial r}{\partial \psi_a} \frac{\partial y}{\partial \psi_r} - \frac{\partial r}{\partial \psi_r} \frac{\partial y}{\partial \psi_a} \right] \end{aligned} \tag{8.42}$$

The second bracket can now be expressed in a better form because, from Eq. (8.32a)

$$\frac{\partial r}{\partial \psi_a} = 0$$
$$\frac{\partial r}{\partial \psi_r} = h \sec \psi_r \tan \psi_r \tag{8.43}$$

and from Eq. (8.32c)

$$\frac{\partial y}{\partial \psi_r} = h \sec^2 \psi_r \sin (\theta_a + \psi_a)$$
$$\frac{\partial y}{\partial \psi_a} = h \tan \psi_r \cos (\theta_a + \psi_a) \tag{8.44}$$

so that the second bracket in Eq. (8.42) can be expressed as

$$\frac{\partial r}{\partial \psi_a} \frac{\partial y}{\partial \psi_r} - \frac{\partial r}{\partial \psi_r} \frac{\partial y}{\partial \psi_a} = -h^2 \sec \psi_r \tan^2 (\psi_r) \cos (\theta_a + \psi_a) \tag{8.45}$$

Finally, by substituting Eq. (8.44), and Eq. (8.45) with the values of r and J from Eq. (8.32a) and Eq. (8.33b) in Eq. (8.42) we obtain

$$
\frac{\partial k(x, y)}{\partial x} = \frac{1}{h^3} \left[\frac{\cos(\theta_a + \psi_a)}{\sec^4 \psi_r} \frac{\partial K(\psi_r, \psi_a)}{\partial \psi_r} - \frac{\sin(\theta_a + \psi_a)}{\sec^2 \psi_r \tan \psi_r} \frac{\partial K(\psi_r, \psi_a)}{\partial \psi_a} \right]
$$
$$
- \frac{2K(\psi_r, \psi_a)}{h^3} \left[\sin \psi_r \cos^3 \psi_r \cos(\theta_a + \psi_a) \right] \tag{8.46}
$$

With this result, the expression of Eq. (8.28) in angular coordinates is

$$
I_3 = \int_{-\pi/2}^{\pi/2} \int_{-\pi/2}^{\pi/2} J \left[\frac{\partial}{\partial x} k(x, y) \right]^2 d\psi_r d\psi_a \tag{8.47}
$$

in which the Jacobian J is given by Eq. (8.33b).

IV. Summary of Derived Expressions

In summary, we have from Eq. (8.23) that the Doppler frequency, which is the displacement δ of the Doppler spectrum center frequency from the transmitter frequency ω_0 is

$$
\delta = \frac{4\pi v}{\lambda} \frac{I_1}{I_0} \tag{8.48}
$$

and from Eq. (8.24) the Doppler spectrum bandwidth σ is

$$
\sigma^2 = v^2 \frac{I_3}{I_0} + \left(\frac{4\pi v}{\lambda} \right)^2 \frac{I_2}{I_0} - \delta^2 \tag{8.49}
$$

in which I_0, I_1, and I_2 are given in angular coordinates by Eqs. (8.34) and I_3 is given in angular coordinates by Eq. (8.47). A specific expression for the spectrum bandwidth is now obtained by substituting Eq. (8.48) in Eq. (8.49)

$$
\sigma^2 = v^2 \frac{I_3}{I_0} + \left(\frac{4\pi v}{\lambda} \right)^2 \left[\frac{I_2 I_0 - I_1^2}{I_0^2} \right] \tag{8.50}
$$

With Eq. (8.48) and Eq. (8.50), the Doppler spectrum quality factor $Q = \delta/2\sigma$ is

$$
\frac{1}{Q^2} = \frac{4I_0}{I_1^2} \left[I_2 + \left(\frac{\lambda}{4\pi} \right)^2 I_3 \right] - 4 \tag{8.51}
$$

These relations allow the determination of the important Doppler spectrum parameters without the need to determine the Doppler spectrum explicitly. With the aid of a computer, these relations enable the study of the effect of the terrain reflectivity $B(\psi_r, \psi_a)$ and the various antenna parameters such as the antenna beamwidth, sidelobe level, and the antenna azimuth and elevation angles upon the Doppler frequency, Doppler spectrum bandwidth, and Doppler spectrum quality factor.

Chapter 9

Computer Study of the Doppler Spectrum

THE EQUATIONS for the center frequency and the bandwidth of the Doppler spectrum derived in Chapter 8 are expressed as the ratio of integrals. Unfortunately, the integrals generally cannot be evaluated analytically. We thus require a computer for the study of the Doppler frequency δ, the Doppler spectrum bandwidth σ, and the Doppler quality factor Q as a function of the expected terrain backscatter, the antenna pattern, and the antenna elevation and azimuth angles.

I. Summary of Equations

We'll first summarize the results obtained in Chapter 8 that are used for our computer study in this chapter.

The derived expression for the Doppler frequency δ is

$$\delta = \omega_c - \omega_0 = \frac{4\pi v}{\lambda}\frac{I_1}{I_0} \text{ rad/s}$$

$$= \frac{2v}{\lambda}\frac{I_1}{I_0} \text{ hertz} \tag{9.1}$$

in which ω_0 is the transmitted frequency and ω_c is the center frequency of the Doppler spectrum. The center frequency ω_c is the frequency at which one-half of the average power of the Doppler spectrum lies above it. The derived expression for the Doppler spectrum bandwidth is

$$\sigma^2 = v^2\frac{I_3}{I_0} + \left(\frac{4\pi v}{\lambda}\right)^2\left[\frac{I_2 I_0 - I_1^2}{I_0^2}\right](\text{rad/sec})^2$$

$$= \frac{v^2}{(2\pi)^2}\frac{I_3}{I_0} + \left(\frac{2v}{\lambda}\right)^2\left[\frac{I_2 I_0 - I_1^2}{I_0^2}\right](\text{hertz})^2 \tag{9.2}$$

129

In the above expressions, the integrals I_n are

$$I_0 = \int_{-\pi/2}^{\pi/2} \int_{-\pi/2}^{\pi/2} \frac{J}{r^4} K^2(\psi_r, \ \psi_a) \, d\psi_r d\psi_a \tag{9.3a}$$

$$I_1 = \int_{-\pi/2}^{\pi/2} \int_{-\pi/2}^{\pi/2} \frac{xJ}{r^5} K^2(\psi_r, \ \psi_a) \, d\psi_r d\psi_a \tag{9.3b}$$

$$I_2 = \int_{-\pi/2}^{\pi/2} \int_{-\pi/2}^{\pi/2} \frac{x^2 J}{r^6} K^2(\psi_r, \ \psi_a) \, d\psi_r d\psi_a \tag{9.3c}$$

$$I_3 = \int_{-\pi/2}^{\pi/2} \int_{-\pi/2}^{\pi/2} J \left[\frac{\partial}{\partial x} k(x, y) \right]^2 d\psi_r d\psi_a \tag{9.3d}$$

In the integrals, J is the Jacobian

$$J = h^2 \sec^2 \psi_r \tan \psi_r \tag{9.4}$$

and

$$\frac{\partial k(x, \ y)}{\partial x} = \frac{1}{h^3} \left[\frac{\cos (\theta_a + \psi_a)}{\sec^4 \psi_r} \frac{\partial K(\psi_r, \ \psi_a)}{\partial \psi_r} - \frac{\sin (\theta_a + \psi_a)}{\sec^2 \psi_r \tan \psi_r} \frac{\partial K(\psi_r, \ \psi_a)}{\partial \psi_a} \right]$$
$$- \frac{2K(\psi_r, \ \psi_a)}{h^3} [\sin \psi_r \cos^3 \psi_r \cos (\theta_a + \psi_a)] \tag{9.5}$$

All the geometric parameters in the equations are shown in Fig. 9.1. The radial distance r from the antenna to a point P on the terrain is

$$r = h \sec \psi_r \tag{9.6}$$

and

$$x = h \tan \psi_r \cos (\theta_a + \psi_a) \tag{9.7}$$

In all these expressions, λ is the wavelength of the radiated sinusoid, v is the antenna velocity, and h is the height of the antenna above the average terrain level. Also, as shown on Fig. 9.1, the center of the antenna pattern on the ground is at the point P_0. The angles are: θ_a is the azimuth angle and θ_e is the elevation angle that the antenna has been rotated; also ψ_r is the range angle and ψ_a is the azimuth angle from the antenna to a point on the terrain.

The effective two-way antenna pattern is

$$K(\psi_r, \ \psi_a) = S(\psi_r, \ \psi_a) B(\psi_r, \ \psi_a) \tag{9.8}$$

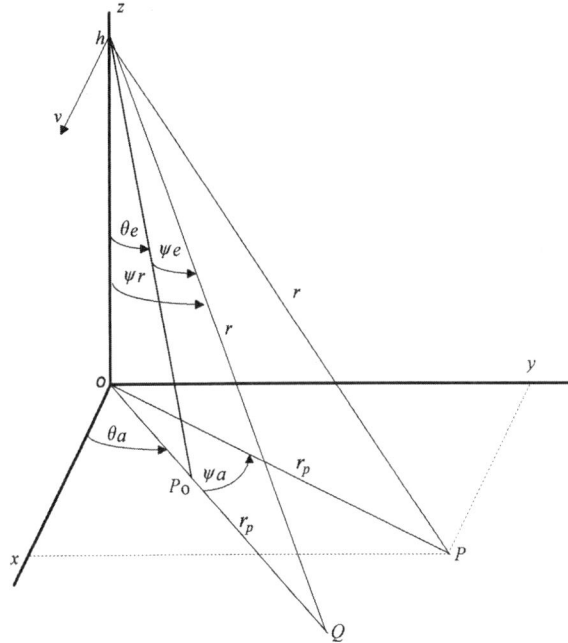

Fig. 9.1   Antenna co-ordinate geometry used for the equations.

in which $S(\psi_r, \psi_a)$ is the two-way antenna pattern and $B(\psi_r, \psi_a)$ is the expected terrain backscatter. For a diffuse surface, the backscatter follows the Lambert law, which is $b \sin \beta$ in which β is the angle between the incident wave and the plane of the terrain. In Fig. 9.1, β is the angle between the $x-y$ plane and the line from the antenna to the point p. Thus, for a terrain that is a diffuse surface, the expected terrain backscatter is

$$B(\psi_r, \psi_a) = b \sin \beta = b \frac{h}{r} \qquad (9.9a)$$

Substituting Eq. (9.6) we obtain that the backscatter function for diffuse terrain is

$$B(\psi_r, \psi_a) = b \cos \psi_r \qquad (9.9b)$$

The theoretical determination of the backscatter for more realistic types of terrain is extremely difficult. However, empirical equations can be obtained from the many experimental studies of the backscatter from various types of terrain that have been reported.

II.   Effect of Height on the Doppler Spectrum

Equation (9.1) and Eq. (9.2) can be put in a more suitable form to observe the effect of the height on the Doppler spectrum. First note that

$$F_1 = \frac{I_1}{I_0} \tag{9.10a}$$

$$F_2 = \frac{I_2}{I_0} \tag{9.10b}$$

and

$$F_3 = \frac{I_3}{I_0} h^2 \tag{9.10c}$$

are invariant with the antenna height h. In terms of these defined parameters, we can express Eq. (9.1) and Eq. (9.2) in the same dimensionless form as used for the quasi-static approximation so that the numerical results are independent of wavelength and aircraft velocity. With the Doppler frequency δ expressed in rad/s, the normalized Doppler frequency is

$$\frac{\lambda}{v}\delta = 4\pi F_1 \tag{9.11a}$$

With δ expressed in hertz, the normalized Doppler frequency is

$$\frac{\lambda}{v}\delta = 2F_1 \tag{9.11b}$$

With the bandwidth σ expressed in rad/s, the normalized bandwidth is

$$\left(\frac{\lambda}{v}\sigma\right)^2 = [BWN]^2 = \left(\frac{\lambda}{h}\right)^2 [BWNL]^2 + (4\pi)^2 [BWNH]^2 \tag{9.12a}$$

With the bandwidth σ expressed in hertz, the normalized bandwidth is

$$\left(\frac{\lambda}{v}\sigma\right)^2 = [BWN]^2 = \left(\frac{\lambda}{2\pi h}\right)^2 [BWNL]^2 + 4[BWNH]^2 \tag{9.12b}$$

in which

$$[BWNL]^2 = F_3 \tag{9.13a}$$

and

$$[BWNH]^2 = F_2 - F_1^2 \tag{9.13b}$$

The Doppler spectrum bandwidth is expressed as the linear combination of two terms, each of which is independent of the antenna height in wavelengths. The quantity $BWNL$ is the normalized low-altitude bandwidth and $BWNH$ is the normalized high-altitude bandwidth. Note that if

$$\left(\frac{BWNL}{BWNH}\right)^2 \ll \left(\frac{4\pi h}{\lambda}\right)^2 \tag{9.14a}$$

then

$$BWN \approx 2\,BWNH \text{ hertz} \tag{9.14b}$$

so that the Doppler spectrum bandwidth is proportional to $BWNH$ if the antenna height in wavelengths is sufficiently large. On the other hand, if

$$\left(\frac{BWNL}{BWNH}\right)^2 \gg \left(\frac{4\pi h}{\lambda}\right)^2 \tag{9.15a}$$

then

$$BWN \approx \left(\frac{\lambda}{2\pi h}\right)BWNL \text{ hertz} \tag{9.15b}$$

so that the Doppler spectrum bandwidth is proportional to $BWNL$ if the antenna height in wavelengths is sufficiently small.

Similarly, the Doppler quality factor $Q = \delta/2\sigma$ can be expressed in the form

$$\frac{1}{Q^2} = \frac{1}{Q_h^2} + \frac{1}{Q_l^2} \tag{9.16}$$

in which

$$Q_h = \frac{F_1}{2(BWNH)} \tag{9.17a}$$

and

$$Q_l = \left(\frac{2\pi h}{\lambda}\right)\frac{F_1}{BWNL} \tag{9.17b}$$

We thus note that $Q \approx Q_h$ if the antenna height in wavelengths is sufficiently large so that Eq. (9.14a) is satisfied; also; $Q \approx Q_l$ if the antenna height in wavelengths is sufficiently small so that Eq. (9.15a) is satisfied.

III.   Illustrative Numerical Study

We'll illustrate the results summarized above by analyzing the case in which the terrain is a diffuse surface for which the expected terrain backscatter is given by Eq. (9.9b) and the two-way antenna pattern is a normal function for which

$$S(\psi_r, \psi_a) = e^{(-\mu_a \psi_e^2 + 2\mu_{ae}\psi_a\psi_e + \mu_e\psi_a^2)/2\sigma_1^2\sigma_2^2} \tag{9.18}$$

in which

$$\mu_a = \sigma_1^2 \sin^2 \phi + \sigma_2^2 \cos^2 \phi \tag{9.19a}$$

$$\mu_e = \sigma_1^2 \cos^2 \phi + \sigma_2^2 \sin^2 \phi \tag{9.19b}$$

$$\mu_{ae} = \left(\sigma_2^2 - \sigma_1^2\right) \sin \phi \cos \phi = \frac{1}{2}\left(\sigma_2^2 - \sigma_1^2\right) \sin 2\phi \tag{9.19c}$$

and, as shown in Fig. 9.1,

$$\psi_e = \psi_r - \theta_e \tag{9.20}$$

The antenna pattern cross-section that is the level curves of $S(\psi_r, \psi_a)$ in the (ψ_e, ψ_a) plane are ellipses in which, as shown in Fig. 9.2, one elliptical axis is at an angle ϕ from the ψ_e axis and has a length proportional to σ_1. The length of the other elliptical axis is proportional to σ_2.

The backscatter and the antenna pattern chosen for our numerical illustration of the exact theory are the same as those used previously for the numerical illustration of the quasi-static approximation in order to use those results as a guide for our study of the exact theory. Also, as discussed in the quasi-static illustration, note that $S(\psi_r, \psi_a)$ is a two-way antenna pattern so that the 3-dB beamwidth is the angle at which $S(\psi_r, \psi_a) = 1/2$. Now $S(\psi_r, \psi_a)$ in Eq. (9.18) is a normal function so that for $\phi = 0$, the 3-dB width along the σ_1 axis is $\psi_e = \sigma_1\sqrt{2\ln2} = 1.17741\sigma_1$ and is $\psi_a = \sigma_2\sqrt{2\ln2} = 1.17741\sigma_2$ along the σ_2 axis. Thus, for example, a two way beamwidth with $\sigma = 3$ degrees corresponds to a two-way 3-dB beamwidth of 3.53 degrees. For our examples, the 3-dB Doppler bandwidth will be proportional to σ given by Eqs. (9.12). However, the proportionality factor is not easily determined because it requires knowledge of the shape of the Doppler spectrum, which is not necessarily a normal function such as that

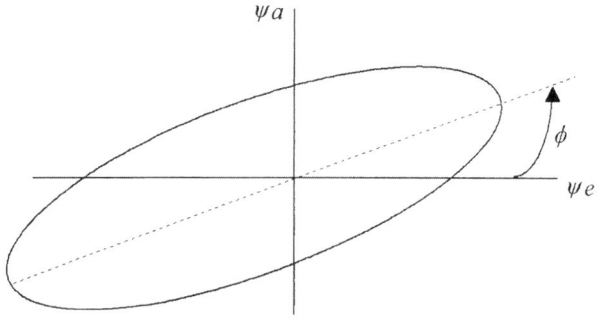

Fig. 9.2   Antenna pattern cross-section level curve.

of the antenna pattern we are using. However, as will be seen from our discussion, the proportionality factor is not really needed.

Before launching into an examination of computer results, a good procedure is to first estimate some of the characteristics of the expected results from physical considerations. If they do not agree with the results obtained from the computer, then either there is an error in the computer program and/or your physical understanding is in error. A number of computer program errors were found in this manner and a better physical understanding of the Doppler spectrum was obtained. A good procedure to follow in all research is to estimate expected results by using your physical understanding before working out the exact theory and doing computations. If the computations do not agree with your expectations, then your theory and/or your physical understanding is in error. Resolving the disagreement often results in a deeper understanding of the phenomenon being studied.

For our present study of the Doppler spectrum, we noted in our discussion of the thin antenna pattern in Chapter 6 that the received Doppler echo is the sum of reflections from many scatterers as they pass through the antenna pattern. The reflection from each scatterer is a pulse that is amplitude modulated and phase modulated. The amplitude modulation is the result of the terrain backscatter and the antenna pattern; the phase modulation is the result of the changing phase of the reflection from the scatterer and the changing radial distance of the scatterer from the radar antenna. The amplitude and phase modulation of the pulses from different scatterers generally will be different. The Doppler spectrum is the spectrum of the waveform, which is the sum of these pulses. The Doppler frequency is the center of this spectrum and the Doppler bandwidth is the bandwidth of this spectrum. The antenna velocity is determined by tracking the Doppler frequency, which is often referred to as the *Doppler shift* although there is no single frequency but a spectrum with a Doppler bandwidth. The tracking accuracy is determined by the Doppler quality factor Q, which is the ratio of the Doppler frequency to the Doppler bandwidth.

The quasi-static approximation of the Doppler spectrum is obtained by ignoring the amplitude modulation of a pulse and assuming that the rate of change of the frequency of each pulse is very small. With these assumptions, the approximation of the Doppler spectrum is thus obtained by assigning a Doppler shift to each point on the terrain in the manner discussed in Chapter 4. Thus we would expect the quasi-static Doppler spectrum to be a good approximation of the exact Doppler spectrum if the effect of the amplitude modulation is small and if the second derivative of the phase (which is the rate of change of the frequency) of each pulse is small.

The equations have been programmed using MatLab 6.3. Before you attempt to run the program using the copy included at the end of this text, you should read Chapter 10, which includes a description of the computer program and its use. The program generates graphs of the quasi-static approximation, my exact theory, their percent difference, and the ratio BWNL/BWNH. Observe that the ratio BWNL/BWNH is the same as the unnormalized ratio BWL/BWH because the same normalizing factor is used in the numerator and the denominator.

The Doppler frequency will be discussed first before the bandwidth and the Q. Similar to the graphs of the quasi-static approximation in Chapter 4, the graphs of the exact Doppler frequency are graphs of the normalized exact Doppler frequency, which is $(\lambda/v)\delta$. The reason for this normalization is that the graphs are then independent of wavelength and aircraft velocity. Furthermore, for convenience in using the graphs, δ is in hertz. The height above the terrain will not be specified because, as can be seen from the equations, both the quasi-static Doppler frequency and the exact Doppler frequency are independent of height. In order to compare the quasi-static approximation with the exact theory, the graph of the normalized Doppler frequency for a given set of parameters will be shown in a group of three: a graph using my exact theory, a graph using the quasi-static approximation, and a graph of the percentage difference, which is defined as

$$\% \text{ Difference} = (100)\left[\frac{(\text{quasi-static approximation}) - (\text{exact theory})}{(\text{exact theory})}\right] \quad (9.21)$$

Figs. 9.3 are graphs of the normalized Doppler frequency versus the elevation angle θ_e for the azimuth angles $\theta_a = 0$, 20, 40, 60, and 80 degrees and for $\sigma_1 = \sigma_2 = 1$ degree two-way antenna beamwidth. Figs. 9.4 are graphs of the normalized Doppler frequency versus the azimuth angle θ_a for the elevation angles $\theta_e = 15$, 30, 45, 60, and 75 degrees and for $\sigma_1 = \sigma_2 = 1$ degree two-way antenna beamwidth. Observe that the percent difference is independent of the

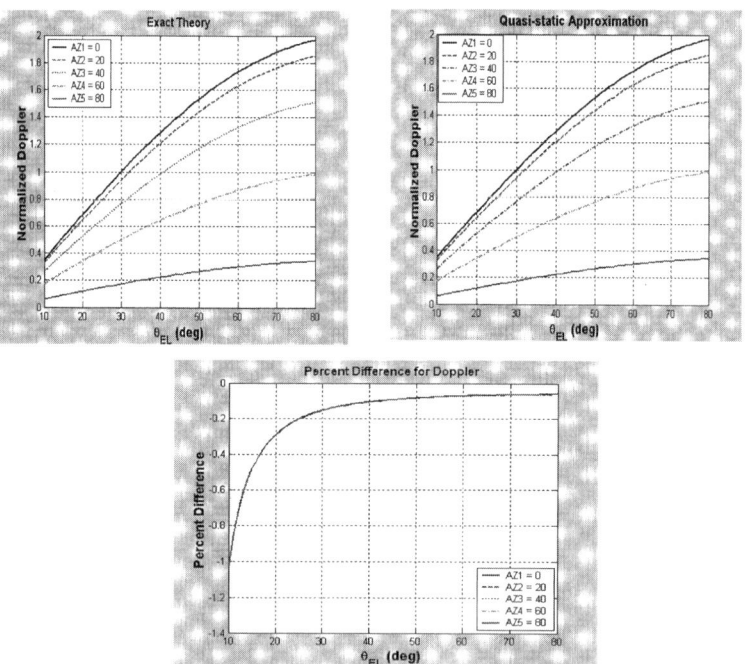

Fig. 9.3   Normalized Doppler frequency versus the elevation angle for $\sigma_1 = \sigma_2 = 1$ degree and for various azimuth angles.

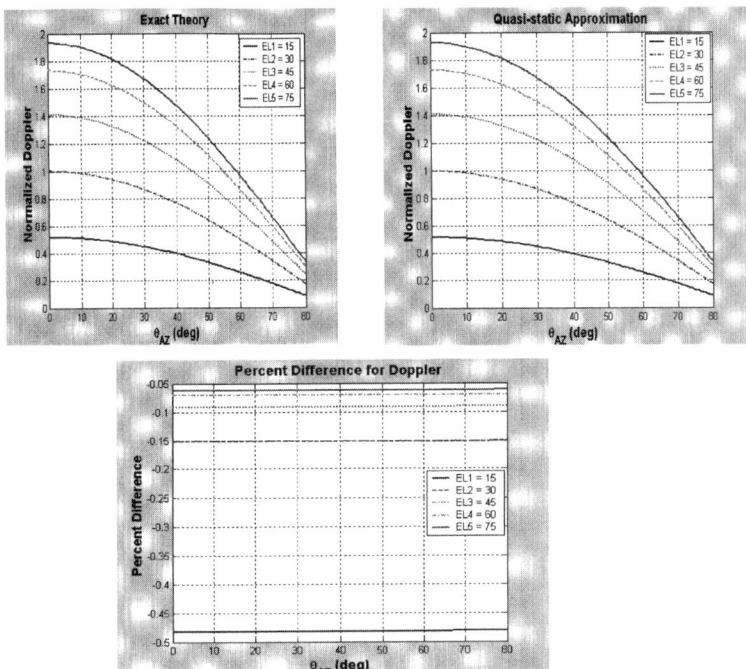

Fig. 9.4   Normalized Doppler frequency versus the azimuth angle for $\sigma_1 = \sigma_2 = 1$ degree and for various elevation angles.

azimuth angle and decreases rapidly for increasing elevation angle. The physical reason is that the rate of change of the frequency of a reflected pulse from a scatterer decreases with increasing elevation angle and the length of the reflected pulse increases with increasing elevation angle so that the percentage error should decrease with increasing elevation angle in accordance with our discussion above.

Figures 9.5 are graphs of the normalized Doppler frequency versus σ_1 for $\sigma_2 = 0.01$, 1.01 degree for $\theta_e = 30$ degrees and $\theta_a = 0$ degrees. The parameter values for Figs. 9.6 are the same except that $\theta_a = 60$ degrees. For both sets of graphs, the rotation angle $\phi = 0$ so that σ_1 is the elevation beamwidth and σ_2 is the azimuth beamwidth. As expected, δ is not a strong function of the azimuth beamwidth σ_2. Observe for both sets of graphs that the graph of the exact theory increases monotonically but the graph of the quasi-static approximation decreases monotonically; the change, however, is small. Also, the graph of the percentage difference is a very weak function of the azimuth angle and the azimuth beamwidth.

Figures 9.7 are graphs of the normalized Doppler frequency versus the rotation angle ϕ for $\sigma_1 = 1.5$ degree, $\sigma_2 = 1$ degree, $\theta_e = 30$ degrees, and $\theta_a = 0$ degrees. The parameter values for Figs. 9.8 are the same except that $\theta_a = 60$ degrees. Note

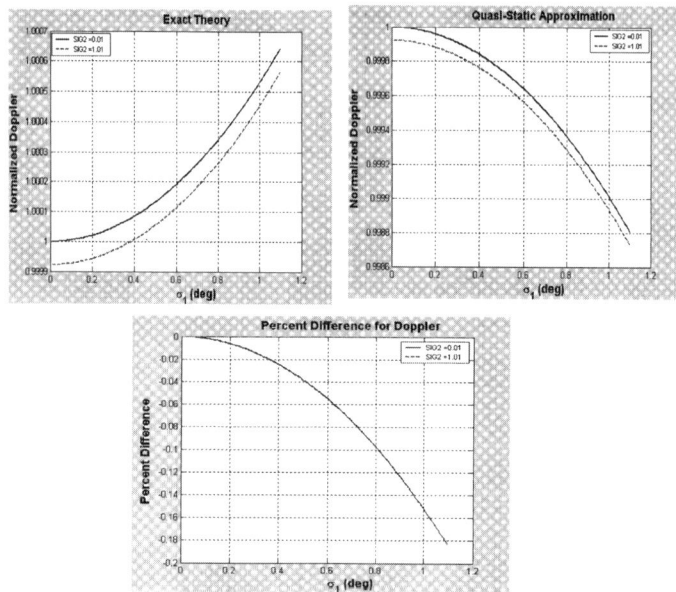

Fig. 9.5 Normalized Doppler frequency versus σ_1 for $\phi = 0$, $\theta_e = 30$ degrees, $\theta_a = 0$, and for various values of σ_2.

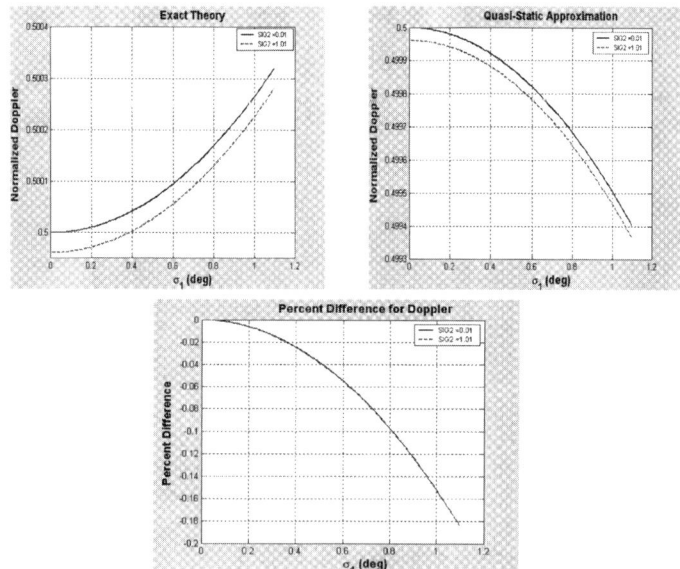

Fig. 9.6 Normalized Doppler frequency versus σ_1 for $\phi = 0$, $\theta_e = 30$ degrees, $\theta_a = 60$ degrees, and for various values of σ_2.

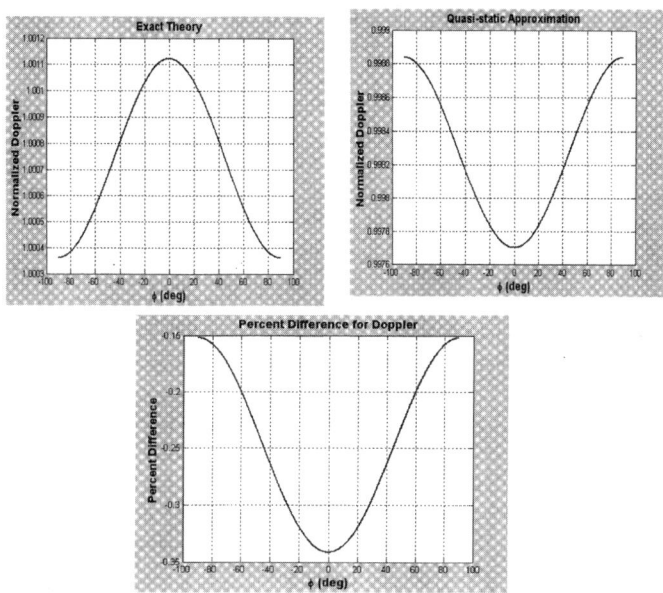

Fig. 9.7 Normalized Doppler frequency versus ϕ for $\theta_e = 30$ degrees, $\theta_a = 0$, $\sigma_1 = 1.5$ degrees, and $\sigma_2 = 1.0$ degree.

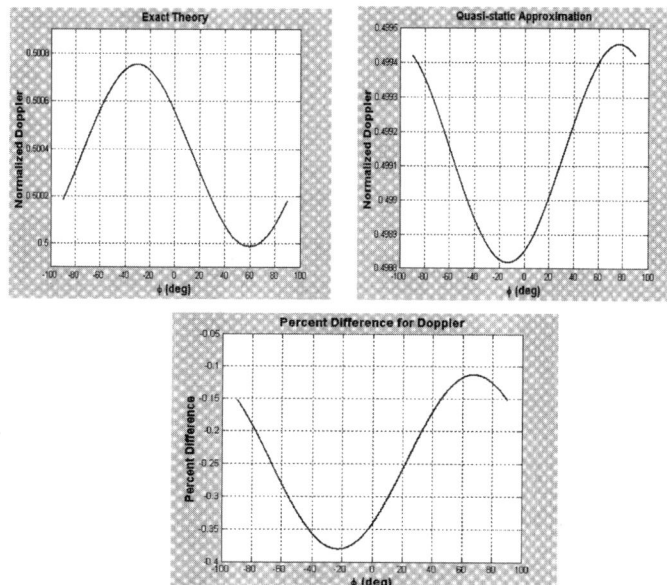

Fig. 9.8 Normalized Doppler frequency versus ϕ for $\theta_e = 30$ degrees, $\theta_a = 60$ degrees, $\sigma_1 = 1.5$ degrees, and $\sigma_2 = 1.0$ degree.

that the exact and the quasi-static Doppler frequency are not strong functions of ϕ as we should expect because they are not strong functions of the beamwidth. However, their variation with rotation angle is significantly different.

In summary, the Doppler frequency (Doppler shift) is independent of altitude. Our numerical study indicates that for elevation angles above about 15 degrees, the percentage error incurred by the quasi-static approximation of the Doppler frequency is less than 1 percent for normal radar antenna beamwidths. Furthermore, from the results in Chapter 4, a good approximation of the normalized quasi-static Doppler frequency $(\lambda/v)\delta_{qs}$ is the normalized frequency assigned to the beam center, $2\sin\theta_e\cos\theta_a$ cycles. This frequency thus is a useful approximation of the exact Doppler frequency.

The Doppler bandwidth however, is another story. First, although the quasi-static approximation of the Doppler bandwidth is independent of height, we observe from Eq. (9.12) that the exact bandwidth is dependent on h/λ, the antenna height in wavelengths above the terrain. This dependence is important because Doppler radar is used on low-flying as well as high-flying vehicles. Also, for normal altitudes, the short wavelength of laser radar can result in very large values of its height in wavelengths. In addition, as opposed to the quasi-static approximation, we showed in Chapter 6 that the exact Doppler bandwidth should increase with decreasing antenna beamwidth as shown in Fig. 6.4. Both of these effects are due to the amplitude modulation of the reflected pulse from each scatterer as we discussed previously. Now, with decreasing antenna beamwidth and with decreasing altitude, the pulse length decreases so that the exact Doppler bandwidth increases. We thus expect the percentage difference between the quasi-static approximation and the exact theory will increase with decreasing antenna beamwidth and decreasing antenna height in wavelengths.

The first set of Doppler bandwidth graphs, Figs. 9.9, are graphs of the normalized quasi-static bandwidth, the normalized exact bandwidth, and their percent difference versus the elevation angle θ_e for $h/\lambda = 10^5$ with antenna azimuth angles $\theta_a = 0$, 20, 40, 60, and 80 degrees with the beamwidth $\sigma_1 = \sigma_2 = 1$ degree. Figures 9.10 are the same set of curves but with an antenna beamwidth $\sigma_1 = \sigma_2 = 0.1$ degree.

Note how different are the shapes of the exact bandwidth curves for the different antenna beamwidths although the shape of the quasi-static approximation curves are the same. In consequence, the shape of the percent difference curves for the two antenna beamwidths are different. Note, too, that as we discussed, the percent difference increases for decreasing antenna elevation angle.

The next set of graphs, Figs. 9.11, are graphs of the normalized quasi-static bandwidth, the normalized exact bandwidth, and their percent difference versus the azimuth angle θ_a for $h/\lambda = 10^5$ with antenna elevation angles $\theta_e = 15$, 30, 45, 60, and 75 degrees with the antenna beamwidth $\sigma_1 = \sigma_2 = 1$ degree. Figures 9.12 are the same set of curves but with an antenna beamwidth $\sigma_1 = \sigma_2 = 0.1$ degree. Again note how different are the shapes of the exact bandwidth curves for the different antenna beamwidths although the shape of the quasi-static approximation curves are the same. In consequence, the shape of the percent difference curves for the two antenna beamwidths are different. Note that the percent

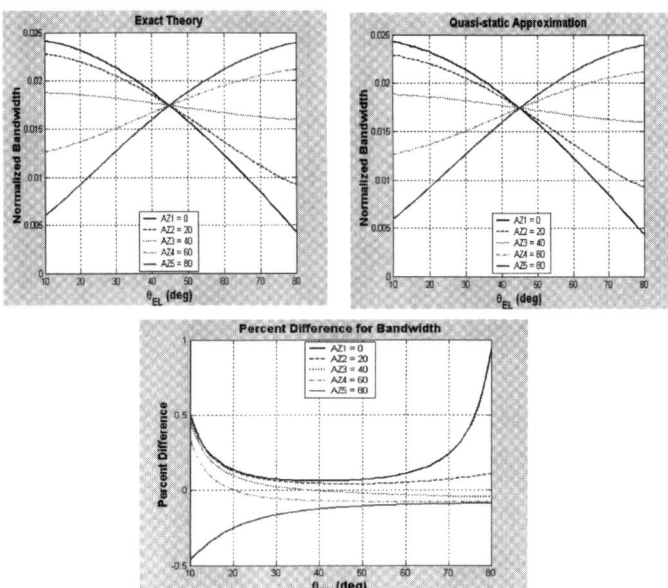

Fig. 9.9   Normalized Doppler bandwidth versus the elevation angle for various azimuth angles with $h/\lambda = 10^5$ and $\sigma_1 = \sigma_2 = 1$ degree.

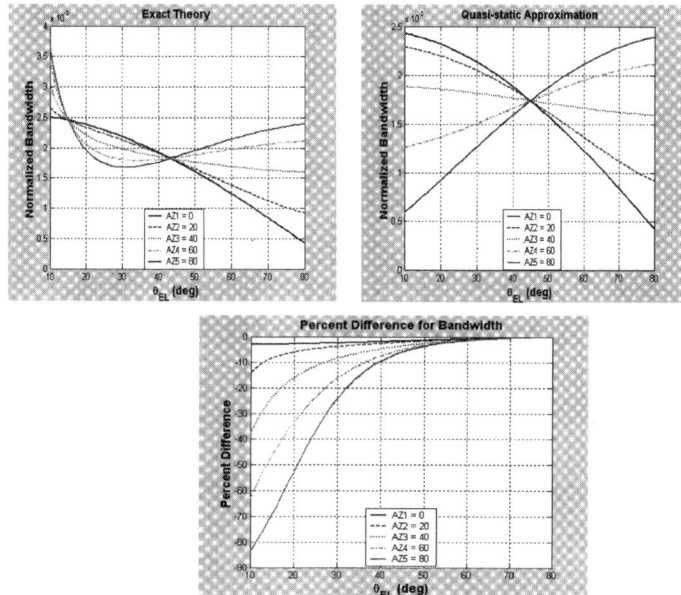

Fig. 9.10   Normalized Doppler bandwidth versus the elevation angle for various azimuth angles with $h/\lambda = 10^5$ and $\sigma_1 = \sigma_2 = 0.1$ degrees.

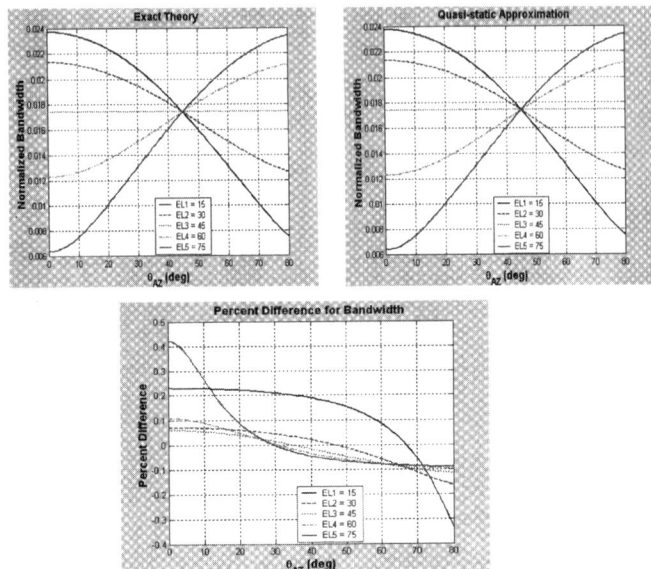

Fig. 9.11 Normalized Doppler bandwidth versus the azimuth angle for various elevation angles with $h/\lambda = 10^5$ and $\sigma_1 = \sigma_2 = 1$ degree.

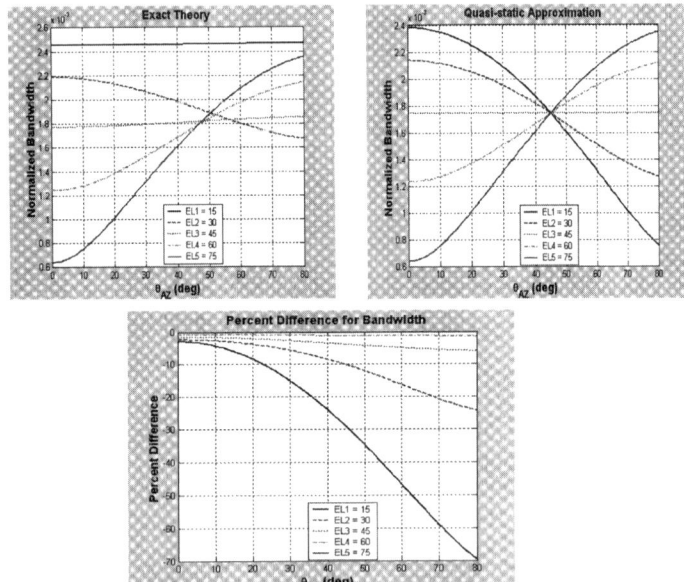

Fig. 9.12 Normalized Doppler bandwidth versus the azimuth angle for various elevation angles with $h/\lambda = 10^5$ and $\sigma_1 = \sigma_2 = 0.1$ degrees.

difference increases for decreasing antenna elevation angles over the whole range of azimuth angles for the narrow beamwidth $\sigma_1 = \sigma_2 = 0.1$ degree but only in the range $25 < \theta_a < 65$ degrees for the for wider antenna beamwidth $\sigma_1 = \sigma_2 = 1.0$ degree. Also, as discussed in Chapter 4, the graphs of the bandwidth versus θ_e or θ_a could be significantly different for $\sigma_1 \neq \sigma_2$ and for $\phi \neq 0$.

Figures 9.13 are graphs of the normalized quasi-static bandwidth, the normalized exact bandwidth, and their percent difference versus σ_1 for $\theta_a = 0$ degrees, $\theta_e = 30$ degrees, $\phi = 0$, and $h/\lambda = 10^5$ for $\sigma_2 = 0.1$ and 1.0 degree. The parameters for the graphs of Figs. 9.14 are the same except for the azimuth angle which is $\theta_a = 30$ degrees. As expected, the bandwidth increases with increasing σ_1 because the larger the beamwidth in the direction of travel, the larger the frequency change over the length of the reflected pulse from each scatterer. Note that as σ_1 decreases, the bandwidth of the quasi-static approximation becomes asymptotic to a value determined by σ_2 and θ_a; the smaller σ_2 or θ_a, the smaller the asymptotic value. Also observe that the exact bandwidth is asymptotic to a curve that is linearly increasing with σ_1. The percentage difference between the quasi-static approximation and the exact theory is small but increases for decreasing values of the antenna beamwidth. The increase is really a result of the bandwidth becoming very small for small values of the beamwidth so that a small difference between the quasi-static approximation and the exact theory is a large percent difference.

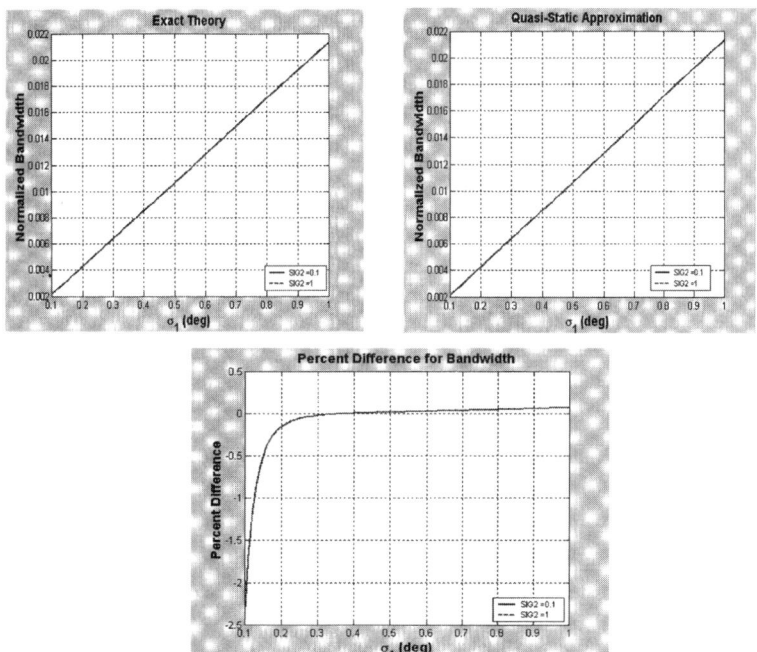

Fig. 9.13 Normalized Doppler bandwidth versus σ_1 for $\sigma_2 = 0.1$ degree and $\sigma_2 = 1.0$ degree with $h/\lambda = 10^5$, $\theta_a = 0$, $\theta_e = 30$ degrees, and $\phi = 0$.

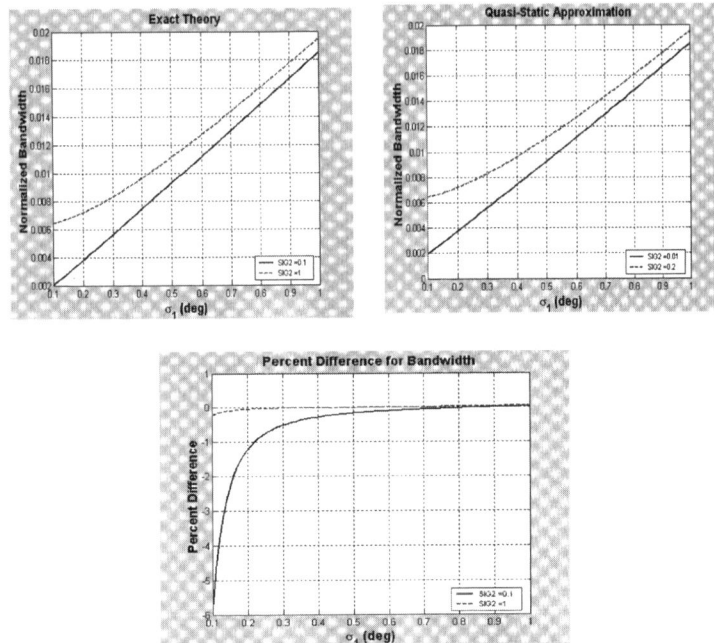

Fig. 9.14   Normalized Doppler bandwidth versus σ_1 for $\sigma_2 = 0.1$ degrees and $\sigma_2 = 1.0$ degree with $h/\lambda = 10^5$, $\theta_a = \theta_e = 30$ degrees, and $\phi = 0$.

It would appear from the graphs that the quasi-static approximation is reasonable for a consideration of the bandwidth. Clearly, this is true for the two cases shown in Figs. 9.13 and Figs. 9.14. However, as opposed to the quasi-static approximation, we discussed that we expect the exact Doppler bandwidth to increase with decreasing antenna beamwidth due to the amplitude modulation of the reflected pulse from each scatterer. This effect is seen in the graphs of Figs. 9.15, which are graphs of the normalized quasi-static bandwidth, the normalized exact bandwidth, and their percent difference versus σ_1 for $\theta_a = 0$ degrees, $\theta_e = 30$ degrees, $\phi = 0$, and $h/\lambda = 10^5$ for $\sigma_2 = 0.01$ and 0.2 degrees. The parameters for the graphs of Figs. 9.16 are the same except for the azimuth angle, which is $\theta_a = 30$ degrees. We note that for both figures, the exact bandwidth is a minimum for $\sigma_1 = 0.048$ degrees so that the quasi-static approximation is clearly not valid for small antenna beamwidths. The two-way beamwidths used in standard x-band radar are normally much larger than this so that we expect the quasi-static approximation to be reasonable for microwave radars. However, the two-way beamwidths used in laser radars can be in this range and so the quasi-static approximation may not be reasonable for laser radars. We'll illustrate this in our discussion concerning the effect of height above the terrain.

The relative insensitivity of the bandwidth with σ_2 for $\phi = 0$, $\theta_a = 30$ degrees, and $\theta_e = 30$ degrees is observed in the graphs of Figs. 9.17. Again observe that for $\sigma_1 = 0.1$ degrees, the increasing percent difference observed for small values

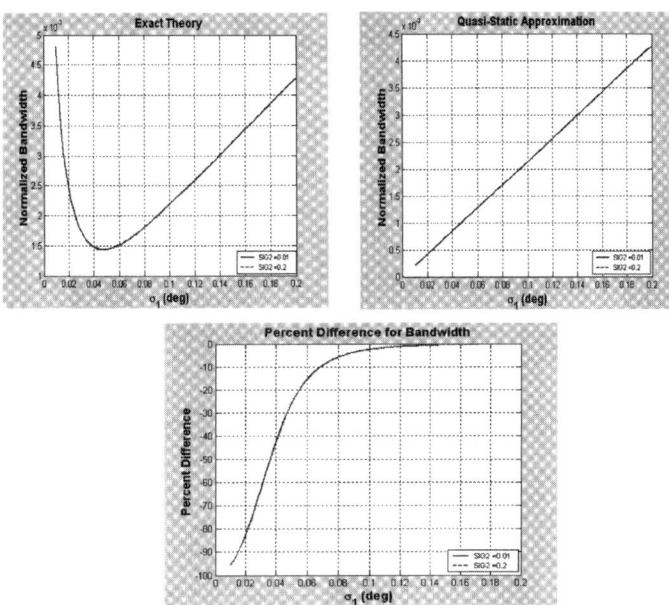

Fig. 9.15   Normalized Doppler bandwidth versus σ_1 for $\sigma_2 = 0.01$ degrees and $\sigma_2 = 0.2$ degrees with $h/\lambda = 10^5$, $\theta_a = 0$, $\theta_e = 30$ degrees, and $\phi = 0$.

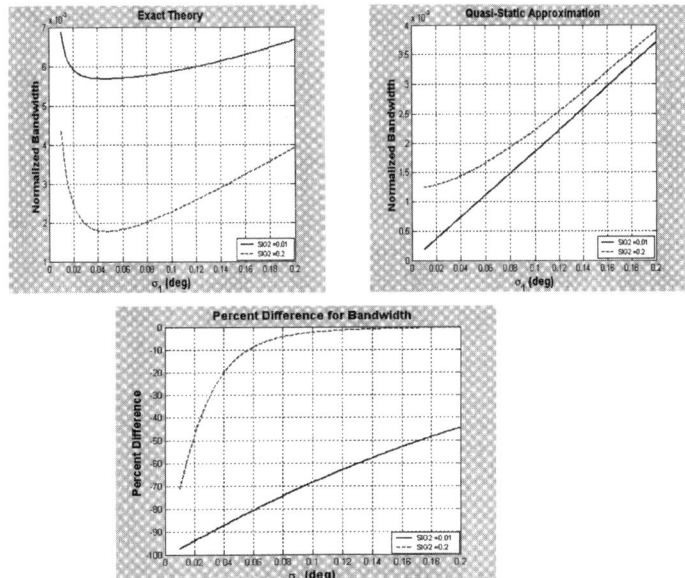

Fig. 9.16   Normalized Doppler bandwidth versus σ_1 for $\sigma_2 = 0.01$ degrees and $\sigma_2 = 0.2$ degrees with $h/\lambda = 10^5$, $\theta_a = \theta_e = 30$ degrees, and $\phi = 0$.

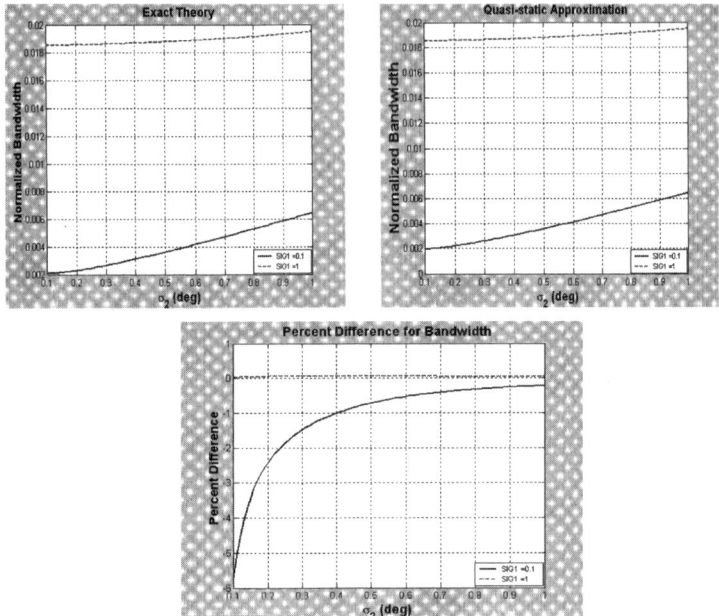

Fig. 9.17   Normalized Doppler bandwidth versus σ_2 for $\sigma_1 = 0.1$ degrees and $\sigma_1 = 1.0$ degrees with $h/\lambda = 10^5$, $\theta_a = \theta_e = 30$ degrees, and $\phi = 0$.

of σ_2 is really a result of the bandwidth becoming very small so that a small difference between the quasi-static approximation and the exact theory is a large percent difference.

We examined the quasi-static bandwidth for an antenna pattern that is not circular in cross-section in Chapter 4. We found that there is an angle of the ellipse for which the bandwidth is a minimum. We found that this angle is the angle at which the major axis of the ellipse lies tangent to an isodoppler given by Eqs. (4.24)

$$\phi_{\min} = \arctan(m) \quad \text{in which} \quad m = \frac{1}{\tan\theta_a \tan\theta_e} \qquad (9.22)$$

This is also a good approximation for large beamwidths because we have observed that the quasi-static approximation is reasonable for large beamwidths. However, because the quasi-static approximation is not reasonable for small beamwidths, we expect Eq. (9.22) not to be valid for small beamwidths.

Figures 9.18 are graphs of the normalized bandwidth versus ϕ, the rotation angle of the ellipse shown in Fig. 9.2 for $\sigma_1 = 0.02$ degrees, $\sigma_2 = 0.01$ degrees, $\theta_a = 30$ degrees, $\theta_e = 30$ degrees, and $h/\lambda = 10^5$. Observe that the minimum for the graph of the quasi-static approximation is 71.6 degrees, which is in agreement with Eq. (9.22) for which $m = 1/\tan 30 \tan 30 = 3$ so that $\phi_{\min} = \arctan(3) = 71.57$ degrees. However, observe from the graph of the exact theory that the angle at which the bandwidth is a minimum for this case is actually

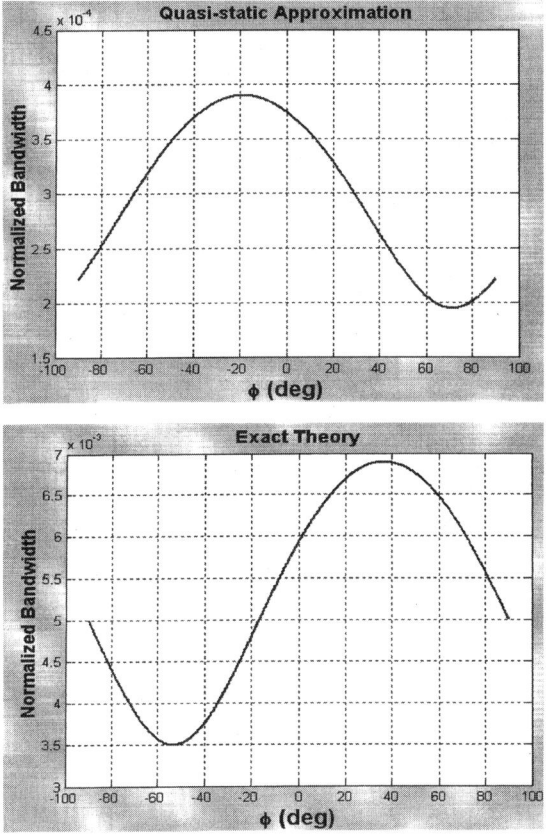

Fig. 9.18   Normalized Doppler bandwidth versus ϕ **for** $h/\lambda = 10^5$, $\sigma_1 = 0.02$ **degree,** $\sigma_2 = 0.01$ **degree, and** $\theta_a = \theta_e = 30$ **degrees.**

-53.2 degrees, which differs from the quasi-static approximation by approximately 125 degrees.

All the graphs above were for an altitude $h/\lambda = 10^5$ in which λ is the wavelength and h is the height of the antenna above the terrain. The quasi-static approximation is independent of the height; however, the exact bandwidth does depend on the height because for a given beamwidth and elevation angle the length of the reflected pulse from each scatterer is proportional to the height of the antenna above the terrain. For example, we observed from Figs. 9.15, that there is a minimum of the exact bandwidth at $\sigma_1 = 0.048$ degrees for a normalized altitude of $h/\lambda = 10^5$. However, if the normalized altitude is reduced to $h/\lambda = 10^4$, then the minimum of the exact bandwidth is increased to $\sigma_1 = 0.15$ degrees and the Doppler Q is $1/3$ its previous value because the exact bandwidth is broader. The lower Q means a lower accuracy with which the Doppler frequency can be determined. The effect of a low altitude on the Doppler Q is thus an important concern for low-flying vehicles, especially those that are designed to fly very low to avoid detection.

IV.   Analysis of a Doppler Laser Radar

The minimum bandwidth occurs at a small antenna beamwidth. Although such small beamwidths are normally not obtained with radars operating at microwave frequencies, they are obtained in short wavelength radars such as laser radars. For example, consider the Firepond Wideband Laser Radar. From its specifications listed in Table 2 of an article describing its design,[1] we find that $\lambda = 11.17 \cdot 10^{-6}$ meters and the one-way 3-dB beamwidth is 10 μrad. In degrees, the one-way 3-dB beamwidth is $5.73 \cdot 10^{-4}$ degrees. For our measure, the corresponding one-way beamwidth is $\sigma = 5.73 \cdot 10^{-4}/1.17741 = 4.9 \cdot 10^{-4}$ degrees. Now, for the given wavelength $h/\lambda = 10^8$ for $h = 1,117$ meter $= 3,667$ feet. Figures 9.19 are graphs of the normalized quasi-static bandwidth and the normalized exact bandwidth for this reasonable altitude $h/\lambda = 10^8$ with

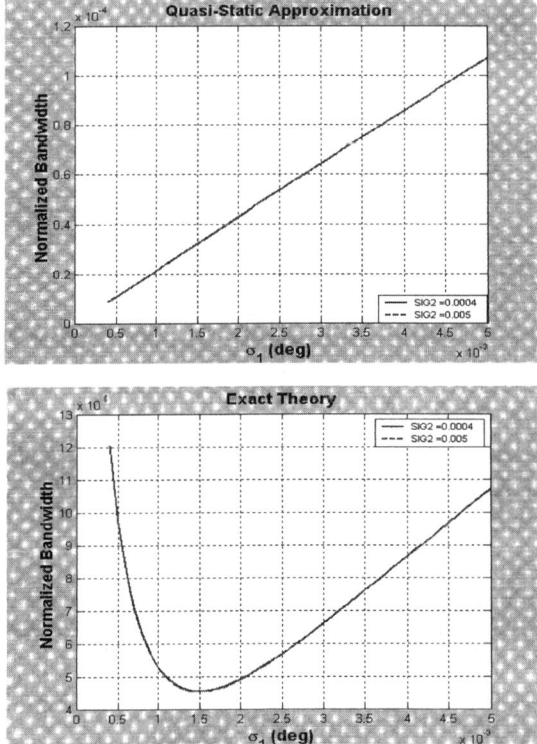

Fig. 9.19   Normalized Doppler bandwidth versus σ_1 for $\sigma_2 = 0.0004$ degrees and $\sigma_2 = 0.005$ degrees with $h/\lambda = 10^8$, $\phi = 0$, $\theta_a = 0$, and $\theta_e = 30$ degrees.

[1]Melngailis et al., "Laser Radar Component Technology," *Proc. IEEE*, vol. 53, No. 2, Feb. 1996, pp. 227–267.

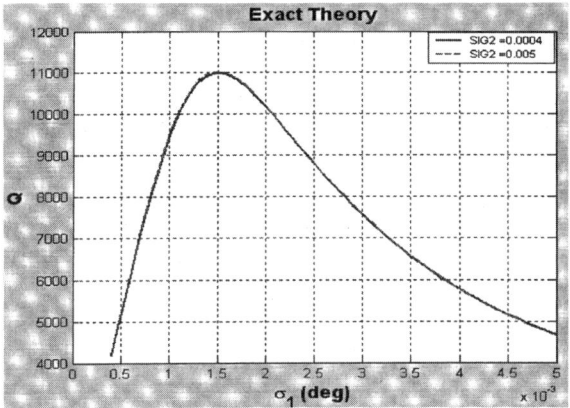

Fig. 9.20 Q versus σ_1 for $\sigma_2 = 0.0004$ degrees, and $\sigma_2 = 0.005$ degrees with $h/\lambda = 10^5$, $\phi = 0$, $\theta_a = 0$, and $\theta_e = 30$ degrees.

$\phi = 0$, $\theta_a = 0$, and $\theta_e = 30$ degrees. As expected, the quasi-static approximation is that the bandwidth decreases with decreasing beamwidth. However, from Figures 9.19, the actual two-way beamwidth has a minimum at $\sigma_1 = 15 \cdot 10^{-4}$ degrees; the corresponding one-way beamwidth is $30 \cdot 10^{-4}$ degrees. Thus, by spoiling the laser radar beam to widen it by a factor of $30 \cdot 10^{-4}/ 4.9 \cdot 10^{-4} = 6.12$, from a one-way 3-dB beamwidth of 10 μrad to a one-way 3-dB beamwidth of 61.2 μrad, we observe from Figures 9.19 that the Doppler bandwidth for this case can be reduced by a factor of approximately $15.6 \cdot 10^{-4}/4.6 \cdot 10^{-4} = 3.4$. The smaller bandwidth means that the Doppler frequency can be determined with less error.[2] Of course, the position accuracy has been reduced because the beamwidth has been increased. This is an example of the Airborne Radar Doppler Uncertainty Principle obtained in Section 6.8A which states that there is a non-zero minimum value of the uncertainty product, $(\Delta v)(\Delta x)$ in which Δv is the uncertainty of velocity and Δx is the uncertainty of position along the direction of the velocity. As discussed in section 6.8A, this uncertainty product is proportional to the product of the Doppler bandwidth and the antenna beamwidth along the direction of the velocity. The tradeoff between the Doppler bandwidth and antenna beamwidth in this example is an illustration of the Airborne Doppler Uncertainty Principle.

---

[2] Note that we have obtained this result for the case in which the terrain is a diffuse surface for which the expected backscatter is given by Eq. (4.17). Even for a smooth sandy beach, this may not be a good approximation because the sand particles are not small relative to the wavelength of a laser. For a theoretical discussion of the scattering of light, see Kerker, M., *The Scattering of Light and Other Electromagnetic Radiation*, Academic Press, New York, NY, 1969 and Beckmann, P., *The Depolarization of Electromagnetic Waves*, Golem Press, Boulder, Colorado, 1968. These texts also contain an extensive list of references. However, backscatter functions are generally determined experimentally because they are not readily determined analytically. For airborne Doppler rader design one should consult references on experimental measurements of backscattering from various types of terrain. The theory presented in this text shows which backscattering parameters are needed.

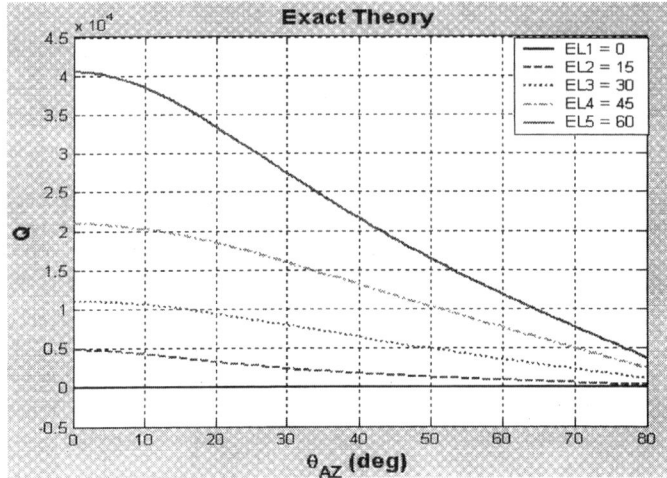

Fig. 9.21 Q versus the azimuth angle for $\theta_e = 0$, 15, 30, 45, and 60 degrees with $h/\lambda = 10^8$, $\phi = 0$, $\theta_a = 0$, and $\theta_e = 30$ degrees.

V.   Measure of Doppler Radar Accuracy

We previously discussed that the accuracy of experimentally determining the Doppler frequency increases with increasing Doppler frequency and decreasing Doppler bandwidth. Thus a measure of the accuracy in experimentally determining the Doppler frequency is the Doppler Q, which I define as

$$Q = \frac{\text{Doppler Frequency}}{\text{Doppler Bandwidth}} \qquad (9.23)$$

Figure 9.20 is a graph of the Q versus σ_1 for the same case as for Fig. 9.19. As expected, the maximum Q is 11,000, which occurs at $\sigma_1 = 15 \cdot 10^{-4}$ degrees. To determine whether the Doppler Q can be made larger at another azimuth or elevation angle, the Q versus the azimuth angle for several elevation angles was obtained and is shown in Fig. 9.21. We observe from the graph that the Doppler Q is a maximum for an antenna pointed in the direction of travel $\theta_a = 0$; however, the Doppler Q increases with increasing elevation angle. Note that the Doppler Q is zero for zero elevation angle because the Doppler frequency is then zero. The Doppler Q increases monotonically with elevation angle because, as we have seen, the Doppler bandwidth decreases with increasing elevation angle. The Doppler Q is not necessarily a maximum at the same parameters at which the Doppler bandwidth is a minimum so that, in design, it is always a good idea to check with graphs of the Doppler Q as we did in this case.

Chapter 10

Computer Program

I. Basic Program Structure

A DISK that contains the MatLab computer program used for the computations presented in this text is enclosed with the text. The antenna pattern
and the backscattering function used by program on the disk are the ones
described in the text. However, other antenna patterns and backscattering functions can be used by modifying the program as described in section 10.3.

The program generates graphs of the quasi-static approximation using the
equations given in section 4.2, my exact theory using the equations given in
section 9.1, their percent difference, and the ratio BWNL/BWNH. Observe
that the ratio BWNL/BWNH is the same as the unnormalized ratio BWL/
BWH because the same normalizing factor is used in the numerator and the
denominator.

MatLab was used for obtaining the graphs. However, MatLab is a vector pro-
cessor so that it is computationally efficient for matrix calculations but rather
inefficient in performing do-loops. Because the computations required extensive
use of do-loops, we have the MatLab data fed to a program written in C++
where all the computations are performed. The results are then fed back to
MatLab for the construction of the graphs.

II. Integration Algorithm

Each of the integrals I_n are integrals over two dimensions. Because an algorithm designed specifically for two dimensions could not be found, I developed
one. For the development of the algorithm, first observe that the integral

$$I = \int_{x_1}^{x_2} \int_{y_1}^{y_2} f(x, y)dx\,dy \tag{10.1}$$

is simply the volume under the surface with height $z = f(x, y)$ over the specified
region of the $x-y$ plane. We shall evaluate this integral by approximating the
surface over a small region by a plane. For this, first consider the solid shown
in Fig 10.1. This is a right triangular solid with a base area equal to A and
bounded above by a slanting plane as shown. Its volume is equal to the base

151

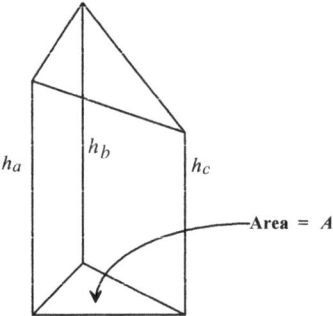

Fig. 10.1   Right triangular solid.

area times the average height of the plane. That is,

$$V = \frac{A}{3}(h_a + h_b + h_c) \tag{10.2}$$

For the evaluation of I in Eq. (10.1), we first divide the x–y plane into right triangles as shown in Fig. 10.2.

Each triangular region has an area equal to $A = (1/2)(\Delta x)(\Delta y)$ in which Δx and Δy are the triangle lengths in the x and y directions, respectively. We then approximate $f(x, y)$ in each triangular region by a plane. The approximate value of I is then the sum of the volumes of all the right triangular solids. With the use of Eq. (10.2), the total volume is

$$V_1 = \frac{A}{3}[f(x_1, \ y_2) + f(x_2, \ y_1) + 2f(x_1, \ y_1) + 2f(x_2, \ y_2)$$

$$+ \ 3 \ \text{times the sum of the values of}$$

$$f(x, y) \ \text{at the other points on the boundry}$$

$$+ \ 6 \ \text{times the sum of the values of}$$

$$f(x, \ y) \ \text{at all the interior points]} \tag{10.3}$$

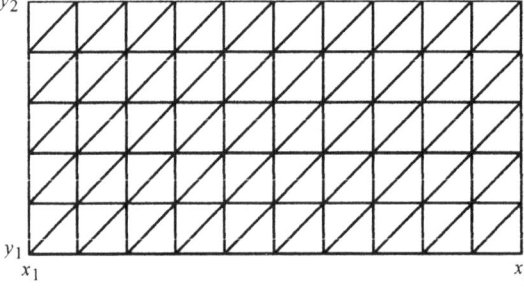

Fig. 10.2   Division of the x–y plane.

Another approximation for the value of I is obtained by forming the triangular regions of the $x-y$ plane as shown in Fig. 10.3. The approximate value of I is again the sum of the volumes of all the right triangular solids. With the use of Eq. (10.2), the total volume is

$$V_2 = \frac{A}{3}[2f(x_1, y_2) + 2f(x_2, y_1) + f(x_1, y_1) + f(x_2, y_2)$$

$+ 3$ times the sum of the values of

$f(x, y)$ at the other points on the boundry

$+ 6$ times the sum of the values of

$f(x, y)$ at all the interior points] $\hspace{2cm}$ (10.4)

The two volumes, V_1 and V_2, are slightly different approximation of I. Wc thus use the average of the two volumes as our approximation of I. That is, our approximation of I is

$$I \approx \frac{1}{2}(V_1 + V_2) \hspace{2cm} (10.5)$$

With the use of Eq. (10.3) and Eq. (10.4), we then have

$$I \approx \frac{A}{2}[f(x_1, y_1) + f(x_1, y_2) + f(x_2, y_1) + f(x_2, y_2)$$

$+ 2$ times the sum of the values of

$f(x, y)$ at the other points on the boundary

$+ 4$ times the sum of the values of

$f(x, y)$ at all the interior points] $\hspace{2cm}$ (10.6)

This is the algorithm used to obtain the approximate value of the double integral in Eq. (10.1).

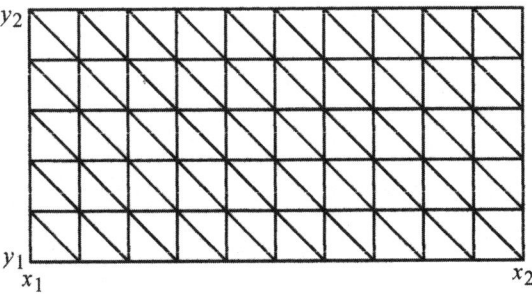

Fig. 10.3 Second division of the $x-y$ plane.

For the program, the integration is over the square which is $\pm 3\sigma_i$ from the beam center. This range was chosen because the error we obtained for this range is less than the round-off error due to the word length. However, this range can be changed by opening the short file *Doppler.hpp* under MatLab, changing the number 3 in

//Integration Parameters

define N_SIGMA   3

to whatever desired, and then recompiling by executing the file makMex.m.

The value of σ_i used is $\max\{\sigma_1, \sigma_2\}$. This choice ensures that the region of integration is sufficiently large for rotations of the antenna pattern ellipse. The integration grid used in the program has equal spacing in the x and y directions so that the rectangles in Fig. 10.2 and Fig. 10.3 are squares.

The length of the sides of each square is chosen to be σ_g in which $[\min\{\sigma_1, \sigma_2\}]/N_g = \sigma_g$ in which N_g is an integer so that there are N_g squares per $\min\{\sigma_1, \sigma_2\}$ along the x and the y directions. The integer value of N_g can be chosen on "Figure No. 1" of the program, which is shown in Fig. 10.4. Also, from beam center to the integration boundary $3\sigma_i$ there are $3\sigma_i/\sigma_g$ intervals. If this is not an integer, the interval size σ_g is deceased slightly to a value

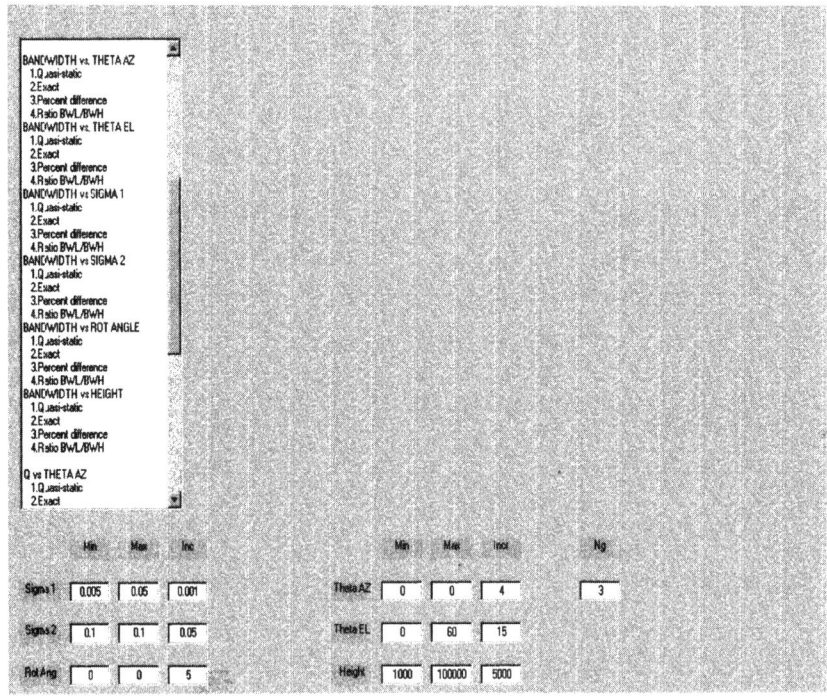

Fig. 10.4   The screen figure, "Figure No. 1."

σ_h so that σ_i/σ_h is the next integer larger than σ_i/σ_g. The integration grid thus always is exactly to the integration boundary.

Because the integration time is proportional to the number of points used in the algorithm, it is proportional to $(\sigma_i/\sigma_h)^2$; this results in a rather large increase of the time for the integration computation for cases in which the ratio of the antenna pattern elliptical axes is large. This program thus can be made faster by choosing the integration region to be a rectangle with sides $3\sigma_1$ and $3\sigma_2$ and rotating the integration grid by an amount equal to the rotation angle ϕ of the antenna ellipse.

III. Using the Program

The program was written using MatLab 6.3. To use the program, version 6.0 or later must be installed. The antenna pattern and the backscattering function used by the program on the disk are the ones described in the text. The program, however, can be modified to use other antenna patterns and back-scattering functions. To define another antenna pattern, open the file antenna_pattern.hpp. The parameters and antenna pattern are defined in a macro called ANT_PATTERN. The parameters are defined in the first part of the macro and the antenna pattern is defined at the end of the macro. Before making changes, save a copy of this macro by making a copy of it as a comment. Then change the antenna pattern to the one desired. The backscatter function can be changed in a similar manner by opening the file back_scatter.hpp and changing the backscatter at the end of the macro BKSCAT to the one desired. After making the desired changes, recompile the program by executing the file makMex.m.

To run the program, first open the file ui.m (ui stands for user interface) on the disk and press F5. On the MatLab Editor Frame that appears, choose "Change MatLab Current Directory" to the one being used. Enlarge the frame "Figure No. 1," which appears to full screen. It will appear as shown in Fig. 10.4.

As discussed previously, the integer value of N_g used can be chosen in the box at the bottom on the right of the screen shown in Fig. 10.4. The default value is 3 because I have obtained very good accuracy with this value. However, this number can be changed to any desired integer.

The left column on the figure lists the various graphs that can be obtained. They are grouped into the following graph types:

1) Doppler frequency versus a parameter
2) Bandwidth versus a parameter
3) Q versus a parameter

For each graph type, the computation can be chosen to be one of the following:

1) Quasi-static approximation
2) Exact theory
3) The percent difference, which is ((quasi-static) − (exact))/(exact) · (100)

The parameter for the abscissa of the graph is chosen from the list of six parameters at the bottom of the screen. The three parameters of the right column are those of the antenna. The first two are the azimuth and elevation angles of the antenna beam axis and the height is the height of the antenna in wavelengths above the average terrain. The three parameters of the left column are those of the antenna pattern ellipse described in Chapter 9. Sigma 1 and Sigma 2 are the lengths of the ellipse axes and the rotation angle is the angle of the Sigma 1 axis relative to the elevation angle. The graph will be plotted from the chosen Min value to the chosen Max value incremented by the chosen Inc value using the max values of the other parameters unless it is a paired value described below.

Several plots of the same graph for different values of a second parameter to which it has been paired can be obtained. The parameters are paired as follows:

1) Sigma 1 and Sigma 2
2) Theta Az and Theta El

Thus on a graph of, say, Q versus Sigma 1, several plots for different values of Sigma 2 can be obtained by choosing the desired Min, Max, and Inc values of Sigma 2. The listing of the various plots will appear in a box on the graph. However, the box can be moved to any desired position by grabbing it with the mouse. Also, the graph can be edited by enabling Plot Editing in the drop-down menu under Tools.

IV.   Brief Description of the Doppler Program File Structure

The following description of the Doppler program file structure is included to aid those who desire to modify the Doppler program as discussed in sections 10.2 and 10.3. This section is also included on the disk as a readme file.

First, there are three files that can be used to implement another antenna pattern and/or a backscattering function and/or set of integration parameter values you desire. They are:

antenna_pattern.hpp

This is an include file for doppler.cpp and qstatic.cpp. It contains a macro in which any desired antenna pattern function can be specified.

back_scattering.hpp

This is an include file for doppler.cpp and qstatic.cpp. It contains a macro in which any desired backscattering function can be specified.

doppler.hpp

This is an include file for doppler.cpp and qstatic.cpp. System parameters and configuration variables such as the size of the integration range in terms of sigma are specified using the #define construct.

The program files are organized in the following hierarchy of levels:

A. Level A

ui.m

> This is a batch file. It clears all global variables, closes all plots, and calls ui_doppler.m.

ui_doppler.m

> This file creates the graphical user interface, Figure No. 1, the plot list menu, and all scalar input fields and initializes all global variables. For this, it calls the file set_test_parameters.m and plot_list.m.

Set_test_parameters.m

> This file reads the values specified in Figure No. 1 and computes the range vectors. The range vectors are global variables. They are used by plot_list.m, mexDispatchEX.m, and mexDispatchQS.m.

plot_list.m

> This file is called by ui_doppler.m. It determines the generation of the plots based on the range and values of the output and the input. It calls for a file in level B.

B. Level B

One of the two files in this level is called depending whether a calculation using the exact or the quasi-static equations is called.

mexDispatchEX.m

> This is a batch file, which is called by plot_list.m. It is used to call the routines for the computation of the exact Doppler spectrum center frequency and bandwidth. It calls for a file in level C1 determined by the request from plot_list.m.

mexDispatchQS.m

> This is a batch file that is called by plot_list.m. It is used to call the routines for the computation of the quasi-static Doppler spectrum center frequency and bandwidth. It calls for a file in level C2 determined by the request from plot_list.m.

C. Level C1

The files in level C1 are for computations using the exact equations. They are called by mexDispatchEX.m.
    dop_azel_ex.cpp

> This is a C++ file that determines the calculations for the Doppler frequency and bandwidth over a range of antenna azimuth and elevation angles. For each pair of azimuth and elevation angle values, it calls Doppler.cpp for the specific computation.

dop_beamwidth_ex.cpp

> This is a C++ file that determines the calculations for the Doppler frequency and bandwidth over a range of antenna beamwidths. For each beamwidth value, it calls doppler.cpp for the specific computation.

dop_rot_ex.cpp

> This is a C++ file that determines the calculations for the Doppler frequency and bandwidth over a range of antenna beam rotation angles. For each rotation angle value, it calls doppler.cpp for the specific computation.

dop_ht_ex.cpp

> This is a C++ file that determines the calculations for the Doppler frequency and bandwidth over a range of antenna heights in wavelengths. For each height, it calls doppler.cpp for the specific computation.

D. Level C2

The files in level C2 are for the computation using the quasi-static equations. They are called by mexDispatchQS.m.
    dop_azel_qs.cpp

> This is a C++ file that determines the calculations for the Doppler frequency and bandwidth over a range of antenna azimuth and elevation angles. For each pair of azimuth and elevation angle values, it calls qstatic.cpp for the specific computation.

dop_beamwidth_qs.cpp

> This is a C++ file that determines the calculations for the Doppler frequency and bandwidth over a range of antenna beamwidths. For each beamwidth value, it calls qstatic.cpp for the specific computation.

dop_rot_qs.cpp

This is a C++ file that determines the calculations for the Doppler frequency and bandwidth over a range of antenna beam rotation angles. For each rotation angle value, it calls qstatic.cpp for the specific computation.

dop_ht_qs.cpp

This is a C++ file that determines the calculations for the Doppler frequency and bandwidth over a range of antenna heights in wavelengths. For each height, it calls qstatic.cpp for the specific computation.

E.   Level D

doppler.cpp

This is a C++ file in which the actual integration of the exact equations is performed for a single set of antenna parameter values called by a file in level C1. The parameter values for the integration algorithm are obtained from doppler.hpp.

qstatic.cpp

This is a C++ file in which the actual integration of the quasi-static equations is performed for a single set of antenna parameter values called by a file in level C2. The parameter values for the integration algorithm are obtained from doppler.hpp.

F.   MISC

makMex.m

This is a Matlab batch file. It is used to compile and link the .cpp files using the Matlab mex compiler.

There are two other file types in the *Doppler* file system. They are the .dll files and the .asv files. The .dll files are executable files. They are the result of compiling and linking the .cpp files. There is one for every .cpp file. The .asv files are backup copies of files edited under the Matlab environment. When a file is edited, a backup .asv file is created from the current unedited version.

The Doppler Spectrum for a Thin Gaussian Antenna Pattern and for $b(x) = b_0$

THE THIN Gaussian antenna pattern given by Eq. (6.69) is

$$s(x) = \frac{1}{a\sqrt{2\pi}} e^{-\left(\frac{(x-x_0)^2}{2a^2}\right)}$$ (A.1)

so that from Eq. (6.19), the complex antenna pattern $p(x)$ is

$$p(x) = \frac{1}{a\sqrt{2\pi}} e^{-\left(\frac{(x-x_0)^2}{2a^2}\right)} e^{-j\left(\frac{4\pi}{\lambda}\right)r}$$ (A.2)

in which $r = \sqrt{h^2 + x^2}$. Then, from Eq. (6.50),

$$P\left(\frac{\omega}{v}\right) = \frac{1}{2\pi} \int_{-\infty}^{\infty} p(x) e^{j(\omega/v)x} dx$$ (A.3)

To analytically evaluate this integral, we require a quadratic approximation for the range r about r_0, the radial distance to the beam center on the ground. For this, we have

$$r = \left[h^2 + x^2\right]^{1/2}$$

First add and subtract x_0^2 and then factor out $r_0^2 = h^2 + x_0^2$

$$= \left[h^2 + x_0^2 + (x^2 - x_0^2)\right]^{1/2}$$

$$= r_0\left[1 + \frac{x^2 - x_0^2}{r_0^2}\right]^{1/2}$$

We now use the approximation

$$\sqrt{1+y} \approx 1 + \tfrac{1}{2}y$$

(for which the error is less than 1 percent for $y < 0.326$) to obtain the approximation

$$r \approx r_0 \left[1 + \frac{x^2 - x_0^2}{2r_0^2} \right] \tag{A.4}$$

Our approximation thus is

$$\frac{x^2 - x_0^2}{r_0^2} < 0.326 \tag{A.5}$$

over the range for which $s(x)$ differs significantly from zero. This range is chosen to be $|x - x_0| < 3a$. Outside this range, $s(x)$ is less than 1.111 percent of its maximum value so the error obtained by using Eq. (A.4) to evaluate the integral Eq. (A.3) is small.

The inequality in Eq. (A.5) can be expressed more meaningfully in terms of the forward-look angle ψ_0 and the angular antenna beamwidth α as shown in Fig. A.1. This is done by noting that the maximum value of $x^2 - x_0^2$ is obtained for $x = x_0 + 3a$ so that

$$\begin{aligned} x^2 - x_0^2|_{\max} &= (x_0 + 3a)^2 - x_0^2 \\ &= 3a[2x_0 + 3a] \end{aligned} \tag{A.6}$$

Thus the maximum value of the ratio in Eq. (A.5) is

$$e = \frac{[x^2 - x_0^2]_{\max}}{r_0^2} = \frac{3a}{r_0} \left[\frac{2x_0}{r_0} + \frac{3a}{r_0} \right] \tag{A.7}$$

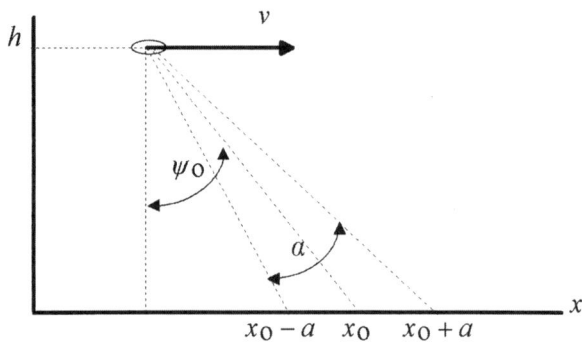

Fig. A.1   Geometry for converting to antenna angular parameters.

Now, from Fig. A.1 we obtain the relations

$$\frac{a}{r_0} = \frac{\alpha}{2\cos\psi_0} \quad \text{and} \quad \frac{x_0}{r_0} = \sin\psi_0 \tag{A.8}$$

Substituting these relations in Eq. (A.7), we then obtain

$$e = \frac{3\sin\psi_0}{\cos\psi_0}\alpha + \left(\frac{3}{2\cos\psi_0}\right)^2\alpha^2 \tag{A.9}$$

For a given value of e, the solution of this quadratic equation for α is

$$\alpha = \frac{2}{3}\left[\sqrt{e + \sin^2\psi_0} - \sin\psi_0\right]\cos\psi_0 \tag{A.10}$$

Figure A.2 is a graph of α versus ψ_0 for $e = 0.32$. Note from this graph that the range of parameters in normal use satisfies this approximation.

We now substitute the approximation Eq. (A.4) in Eq. (A.2) to obtain

$$P\left(\frac{\omega}{v}\right) = \frac{1}{2\pi a}\frac{1}{\sqrt{2\pi}}\int_{-\infty}^{\infty} e^{-((x-x_0)^2/2a^2)}e^{-j(4\pi/\lambda)[r_0 + (x^2 - x_0^2)/(2r_0)]}e^{j(\omega/v)x}dx \tag{A.11}$$

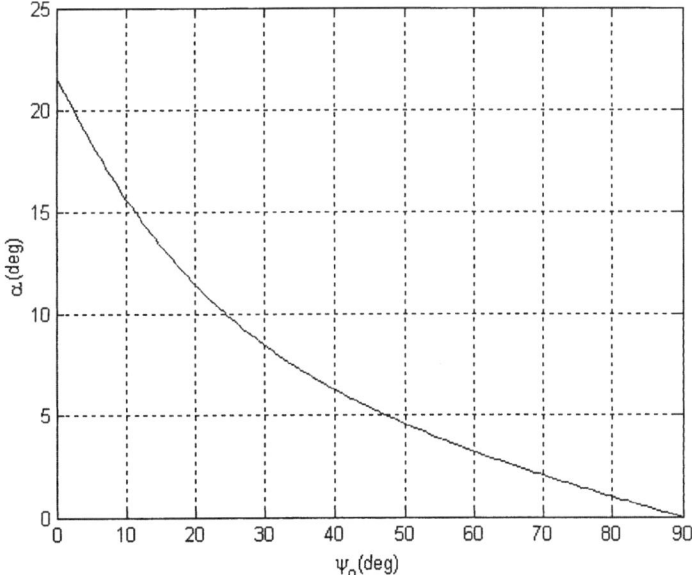

Fig. A.2   Graph of α vs ψ_0 for $e = 0.32$.

With the change of variable $z = x - x_0$, this integral can be expressed as

$$P\left(\frac{\omega}{v}\right) = \frac{1}{2\pi a\sqrt{2\pi}}e^{-j(4\pi/\lambda)r_0}e^{j(\omega/v)x_0}\int_{-\infty}^{\infty}e^{-[(1/2a^2)+j(2\pi/\lambda r_0)]z^2+[(\omega/v)-(4\pi/\lambda)(x_0/r_0)]z}dz$$

$$= \frac{1}{2\pi a\sqrt{2\pi}}e^{-j(4\pi/\lambda)r_0}e^{j(\omega/v)x_0}F\left(\frac{\omega-\omega_d}{v}\right) \tag{A.12}$$

in which

$$F(\omega) = \int_{-\infty}^{\infty}e^{-[(1/2a^2)+j(2\pi/\lambda r_0)]z^2}e^{j\omega z}\,dz \tag{A.13}$$

and

$$\omega_d = \frac{4\pi v}{\lambda}\frac{x_0}{r_0} = \frac{4\pi v}{\lambda}\sin\psi_0 \tag{A.14}$$

Note from Eq. (A.12) that the center of $|P(\omega/v)|$ is displaced by an amount equal to the usual Doppler frequency. Now, an analytic expression for Eq. (A.13) can be obtained from the identity

$$\int_{-\infty}^{\infty}e^{-cz^2}e^{j\omega z}\,dz = \sqrt{\frac{\pi}{c}}\,e^{-(\omega^2/4c)}; \quad \text{Re}\{c\} > 0 \tag{A.15}$$

The result is

$$F(\omega) = \sqrt{\frac{\pi}{1/(2a^2)+j(2\pi/\lambda r_0)}}\,e^{-(\omega^2/(4[(1/(2a^2))+j(2\pi/\lambda r_0)]))} \tag{A.16}$$

With the use of the relation

$$\frac{1}{(1/(2a^2))+j(2\pi/\lambda r_0)} = \frac{(1/(2a^2))-j(2\pi/\lambda r_0)}{(1/2a^2)^2+(2\pi/\lambda r_0)^2} \tag{A.17}$$

and with the substitution of Eq. (A.16) in Eq. (A.12), we obtain

$$P\left(\frac{\omega}{v}\right) = \frac{1}{2\pi\sqrt{1+j(4\pi a^2/\lambda r_0)}}e^{-j(4\pi r_0/\lambda)}e^{j(\omega x_0/v)}e^{-((\omega-\omega_d^2)/(4v^2[1/2a^2+j(2\pi/\lambda r_0)]))}$$

$$\tag{A.18}$$

With the rationalization given by Eq. (A.17), we can express Eq. (A.18) as

$$P\left(\frac{\omega}{\nu}\right) = a\frac{1}{2\pi\sqrt{1 + j(4\pi a^2/\lambda r_0)}}$$

$$\times e^{\left(\begin{array}{c}-((\omega-\omega_d)^2/(8a^2\nu^2[(1/2a^2)^2+(2\pi/\lambda r_0)^2]))-j[((4\pi r_0)/\lambda)\\ -((\omega x_0)/\nu)+(2\pi(\omega-\omega_d)^2/(4\lambda r_0\nu^2[(1/2a^2)^2+(2\pi/\lambda r_0)^2]))]\end{array}\right)}$$

(A.19)

For constant terrain parameter $b(x) = b_0$, we have from Eq. (6.48)

$$\left|G\left(-\frac{\omega}{\nu}\right)\right|^2 = b_0^2\left|P\left(\frac{\omega}{\nu}\right)\right|^2$$

$$= \frac{b_0^2}{(2\pi)^2\sqrt{1 + (4\pi a^2/\lambda r_0)^2}}\, e^{-((\omega-\omega_d)^2/2\sigma^2)}$$

(A.20)

in which we have defined

$$2\sigma^2 = 4a^2\nu^2\left[\left(\frac{1}{2a^2}\right)^2 + \left(\frac{2\pi}{\lambda r_0}\right)^2\right]$$

$$= \left(\frac{\nu}{a}\right)^2\left[1 + \left(\frac{4\pi a^2}{\lambda r_0}\right)^2\right]$$

(A.21)

The Effect of the Terrain
Parameter $b(x)$

$\mathbf{F}$OR THE thin Gaussian pattern, we can obtain a better analytic approximation of the effect of the terrain parameter $b(x)$ upon the echo spectrum than was obtained for the narrow antenna pattern obtained in section 6.7B. For our approximation, we consider the case in which there is a linear variation of the parameter $b(x)$ over the central region of the antenna pattern.

The basic equation required is obtained by substituting Eq. (A.18) in Eq. (6.63). For this, we have from Eq. (A.17) and (A.18) that

$$P(\omega) = Ae^{j\omega x_0}e^{-(\omega-(\omega_d/v))^2/4dv^2} \tag{B.1}$$

in which

$$A = \frac{1}{2\pi\sqrt{1+j(4\pi a^2/\lambda r_0)}}e^{-j(4\pi r_0/\lambda)} \tag{B.2}$$

and

$$d = \frac{1}{2a^2} + j\frac{2\pi}{\lambda r_0} \tag{B.3}$$

Then

$$\frac{d}{d\omega}P(\omega) = \left[jx_0 - \frac{1}{2d}\left(\omega - \frac{\omega_d}{v}\right)\right]P(\omega) \tag{B.4}$$

so that, from Eq. (6.63),

$$G(-\omega) = \left[b_0 - j\frac{m}{2d}\left(\omega - \frac{\omega_d}{v}\right)\right]P(\omega) \tag{B.5}$$

and

$$\left| G\left(-\frac{\omega}{v} \right) \right|^2 = \left| B\left(\frac{\omega}{v} \right) \right|^2 \left| P\left(\frac{\omega}{v} \right) \right|^2 \tag{B.6}$$

in which we have defined

$$B\left(\frac{\omega}{v} \right) = b_0 - j\frac{m}{2dv}(\omega - \omega_d) \tag{B.7}$$

Substituting Eq. (B.3) in Eq. (B.7)

$$\begin{aligned} B\left(\frac{\omega}{v} \right) &= b_0 - j\frac{m}{2v}\frac{(\omega - \omega_d)}{(1/2a^2) + j(2\pi/\lambda r_0)} \\ &= \left[b_0 - \frac{2\pi m a^2 v}{\lambda r_0 \sigma^2}(\omega - \omega_d) \right] - j\frac{mv}{2\sigma^2}(\omega - \omega_d) \end{aligned} \tag{B.8}$$

in which

$$2\sigma^2 = \left(\frac{v}{a} \right)^2 \left[1 + \left(\frac{4\pi a^2}{\lambda r_0} \right)^2 \right] \tag{B.9}$$

which is the same as given by Eq. (A.21). Thus

$$\left| B\left(\frac{\omega}{v} \right) \right|^2 = \left[b_0 - \frac{2\pi m a^2 v}{\lambda r_0 \sigma^2}(\omega - \omega_d) \right]^2 + \left[\frac{mv}{2\sigma^2}(\omega - \omega_d) \right]^2 \tag{B.10}$$

For convenience, we express this equation in the form

$$\left| B\left(\frac{\omega}{v} \right) \right|^2 = b_0^2 - C(\omega - \omega_d) + D(\omega - \omega_d)^2 \tag{B.11}$$

in which

$$C = \frac{2(2\pi)ma^2 v}{\lambda r_0 \sigma^2} \tag{B.12}$$

and

$$D = \frac{m^2 v^2}{4\sigma^2} \left[1 + \frac{4(2\pi)^2 a^4}{\lambda^2 r_0^2} \right] \tag{B.13}$$

We then obtain by substituting Eq. (B.11) in Eq. (B.6) and using Eq. (B.1),

$$\left|G\left(-\frac{\omega}{v}\right)\right|^2 = |A|^2\left[b_0^2 - C(\omega - \omega_d) + D(\omega - \omega_d)^2\right]e^{-(\omega-\omega_d)^2/2\sigma^2} \quad (B.14)$$

To determine the velocity, a Doppler radar tracks the center frequency of the echo spectrum ω_c. This is the frequency below which lies one-half of the total echo spectrum power. The center frequency thus is that frequency ω_c for which

$$\int_{-\infty}^{\omega_c}\left|G\left(-\frac{\omega}{v}\right)\right|^2 d\omega = \int_{\omega_c}^{\infty}\left|G\left(-\frac{\omega}{v}\right)\right|^2 d\omega \quad (B.15)$$

To determine ω_c for our present case, we substitute Eq. (B.14) in Eq. (B.15) and then let

$$\omega_c = \omega_d - \delta \quad \text{and} \quad \omega - \omega_d = x \quad (B.16)$$

to obtain

$$\int_{-\infty}^{-\delta}[b_0^2 - Cx + Dx^2]e^{-(x^2/2\sigma^2)}dx = \int_{-\delta}^{\infty}[b_0^2 - Cx + Dx^2]e^{-(x^2/2\sigma^2)}\,dx \quad (B.17)$$

To determine δ from this equation, we express it in a more convenient form by first adding to both sides of this equation the integral from $-\delta$ to δ to obtain

$$\int_0^{-\delta}[b_0^2 - Cx + Dx^2]e^{-(x^2/2\sigma^2)}\,dx = \int_0^{\infty}[b_0^2 - Cx + Dx^2]e^{-(x^2/2\sigma^2)}\,dx$$
$$- \int_{-\infty}^{0}[b_0^2 - Cx + Dx^2]e^{-(x^2/2\sigma^2)}dx \quad (B.18)$$

Then, making use of the even and odd parts of the integrand, we obtain

$$\int_0^{\delta}[b_0^2 + Cx + Dx^2]e^{-x^2/2\sigma^2}\,dx = \int_0^{\infty}Cxe^{-(x^2/2\sigma^2)}dx \quad (B.19)$$

The value of the integral on the right side of this equation is $C\sigma^2$. Thus, we need to determine the value of δ for which

$$\int_0^{\delta}[b_0^2 + Cx + Dx^2]e^{-(x^2/2\sigma^2)}dx = C\sigma^2 \quad (B.20)$$

This is a transcendental equation for δ. We shall obtain an approximate value of the integral for the usual case in which the shift δ from the Doppler frequency ω_d is small compared with the bandwidth of the echo spectrum, that is, for the case in which

$$\frac{\delta}{\sigma\sqrt{2}} \ll 1 \qquad (B.21)$$

For such a case, $e^{-x^2/2\sigma^2} \approx 1$ over the range of integration so that we have from Eq. (B.20)

$$b_0^2\delta + \frac{C\delta^2}{2} + \frac{D\delta^3}{3} = C\sigma^2 \qquad (B.22)$$

This cubic in δ can be solved exactly. However, with the inequality of Eq. (B.21), we shall take as our approximate solution

$$b_0^2\delta \approx C\sigma^2 \qquad (B.23)$$

Substituting the value of C from Eq. (B.12), we have

$$\delta \approx \frac{2(2\pi)ma^2v}{\lambda r_0 b_0} \qquad (B.24)$$

Normally m defined by Eq. (6.60) is positive so that δ in Eq. (B.24) is positive. We thus obtain from Eq. (B.16) that the shift of the center of the spectrum is to a frequency lower than the Doppler frequency ω_d. Now, from Eqs. (6.59), the Doppler frequency is

$$\omega_d = \frac{2(2\pi)vx_0}{\lambda r_0} \qquad (B.25)$$

so that

$$\frac{\delta}{\omega_d} \approx \frac{ma^2}{b_0x_0} \qquad (B.26)$$

Derivation of Parseval Relations

LET $f_1(x, y)$ and $f_2(x, y)$ be two complex valued functions in L_1 so that for $a = 1$ or 2

$$\int_{-\infty}^{\infty} \int_{-\infty}^{\infty} |fa(x, y)| dxdy < \infty \tag{C.1}$$

This condition is sufficient to ensure the convergence of all our integrals. For $a = 1$ or 2, the two-dimensional Fourier transform pair of these functions is from Eqs. (5.31).[1]

$$Fa(\omega_1, \omega_2) = \frac{1}{(2\pi)^2} \int_{-\infty}^{\infty} \int_{-\infty}^{\infty} fa(x, y) e^{j(\omega_1 x + \omega_2 y)} dxdy \tag{C.2}$$

$$fa(x, y) = \int_{-\infty}^{\infty} \int_{-\infty}^{\infty} Fa(\omega_1, \omega_2) e^{-j(\omega_1 x + \omega_1 y)} d\omega_1 d\omega_2 \tag{C.3}$$

The functions and their transforms then satisfy the Parseval relation

$$\int_{-\infty}^{\infty} \int_{-\infty}^{\infty} F_1(\omega_1, \omega_2) F_2^*(\omega_1, \omega_2) d\omega_1 d\omega_2$$

$$= \frac{1}{(2\pi)^2} \int_{-\infty}^{\infty} \int_{-\infty}^{\infty} f_1(x, y) f_2^*(x, y) dxdy \tag{C.4}$$

We now note from Eq. (8.8) that

$$G(-\omega_1, -\omega_2) = \frac{1}{(2\pi)^2} \int_{-\infty}^{\infty} \int_{-\infty}^{\infty} g(x, y) e^{j(\omega_1 x + \omega_2 y)} dxdy \tag{C.5}$$

---

[1] For Eq. (A.2) and Eq. (A.3) to be a Fourier transform pair, it is just required that the product of the constants before each integral be $1/(2\pi)^2$. We use the factor 1 before the integral in Eq. (A.3) and the factor $1/(2\pi)^2$ before the integral in Eq. (A.2) for consistency with our discussion of the Doppler power density spectrum.

in which $g(x, y)$ is in L_1. Then, with $f_1(x, y) = f_2(x, y) = g(x, y)$ in Eq. (C.4), we have

$$\int_{-\infty}^{\infty}\int_{-\infty}^{\infty} |G(-\omega_1, -\omega_2)|^2\, d\omega_1 d\omega_2 = \frac{1}{(2\pi)^2}\int_{-\infty}^{\infty}\int_{-\infty}^{\infty} |g(x, y)|^2\, dxdy \qquad \text{(C.6)}$$

Further, let

$$fx(x, y) = \frac{\partial}{\partial x} f(x, y) \qquad \text{(C.7)}$$

in which both $f(x, y)$ and $f_x(x, y)$ are in L_1. Then the Fourier transform of $f_x(x, y)$ exists and is

$$Fx(\omega_1, \omega_2) = -j\omega_1 F(\omega_1, \omega_2) \qquad \text{(C.8)}$$

in which $F(\omega_1, \omega_2)$ is the Fourier transform of $f(x, y)$. Consequently, by letting $f_1(x, y) = f_2(x, y) = \frac{\partial}{\partial x} g(x, y)$ in Eq. (C.4), we have

$$\int_{-\infty}^{\infty}\int_{-\infty}^{\infty} \omega_1^2 |G(-\omega_1, -\omega_2)|^2\, d\omega_1 d\omega_2 = \frac{1}{(2\pi)^2}\int_{-\infty}^{\infty}\int_{-\infty}^{\infty} \left|\frac{\partial}{\partial x} g(x, y)\right|^2\, dxdy \quad \text{(C.9)}$$

Finally, by letting $f_1(x, y) = g(x, y)$ and $f_2(x, y) = \partial/\partial x\, g(x, y)$ in Eq. (C.4) we have

$$\int_{-\infty}^{\infty}\int_{-\infty}^{\infty} \omega_1 |G(-\omega_1, -\omega_2)|^2\, d\omega_1 d\omega_2 = \frac{j}{(2\pi)^2}\int_{-\infty}^{\infty}\int_{-\infty}^{\infty} g^*(x, y)\frac{\partial}{\partial x} g(x, y)\, dxdy$$

$$\text{(C.10)}$$

Index

Supporting Materials

A complete listing of titles in Progress in Astronautics and Aeronautics and other AIAA publications is available at http://www.aiaa.org.

To download your software and any software updates, please go to http://www.aiaa.org/publications/supportmaterials. Follow the instructions provided and enter the following password: **antenna**.